Lecture Notes in Mathematics

Edited by A. Dold and B. Eckmann

1278

S. S. Koh (Ed.)

Invariant Theory

Springer-Verlag

Berlin Heidelberg New York London Paris Tokyo

Editor

Sebastian S. Koh
Department of Mathematical Sciences, West Chester University
West Chester, PA 19383, USA

Mathematics Subject Classification (1980): 20-G

ISBN 3-540-18360-4 Springer-Verlag Berlin Heidelberg New York
ISBN 0-387-18360-4 Springer-Verlag New York Berlin Heidelberg

Printed in Germany

Printing and binding: Druckhaus Beltz, Hemsbach/Bergstr.
2146/3140-543210

FOREWORD

This symposium is an outgrowth of a conference on Combinatorics and Invariant Theory held at West Chester University in the Summer of 1985. We felt at that time that a collection of well placed expository papers leading directly to the heart of current research in Invariant Theory would serve an useful purpose. We hope that the present volume has, in some measure, achieved that aim.

We would like to thank Dr. Frank Grosshans for his informative INTRODUCTION which unifies the individual papers into an organized whole.

Editor

LIST OF PAPERS

INTRODUCTION

Invariant theory was developed in the nineteenth century by Boole, Cayley, Clebsch, Gordan, Hilbert, Sylvester, and others. It has been studied intermittently ever since. In recent times, newly developed techniques from algebraic geometry and combinatorics have been applied with great success to some of its outstanding problems. This has moved invariant theory, once again, to the forefront of mathematical research.

In this introduction, we shall introduce the main problems of invariant theory and show how the papers in this collection are related to them. We begin with the necessary definitions.

Let K be an infinite field. Let V be a finite-dimensional vector space over K and let $\{e_1,\dots,e_n\}$ be any basis for V. Let K $[x_1,\dots,x_n]$ be the polynomial algebra in n variables over K. A function $f:V\longrightarrow K$ is called a <u>polynomial function</u> on V if there is a polynomial p in K $[x_1,\dots,x_n]$ such that for every v in V, $v = \sum_{i=1}^{n} a_i e_i$, we have $f(v) = p(a_1,\dots,a_n)$. We shall denote the algebra of all polynomial functions on V by K[V].

Next, let G be a group and suppose that G acts on V via some representation. We say that a polynomial function f on V is <u>invariant</u> with respect to G if $f(g\cdot v) = f(v)$ for all g in G and v in V. The set of all such invariant polynomials forms an algebra which we shall denote by $K[V]^G$.

Invariant theory covers a wide range of highly specialized problems and techniques. One way to impose some unity on this study is to use the notion of an orbit. Let v be any point in V; the <u>orbit</u> of v with respect to the action of G consists of all the points $g\cdot v$ where g is any element of G. We shall denote this orbit by $G\cdot v$. By definition, a polynomial f in $k[V]^G$ is constant on any orbit. The orbit $G\cdot v$ is called <u>separated</u> if $G\cdot v = \{v'\in V: f(v') = f(v) \text{ for all } f\in K[V]^G\}$. In general, not all orbits are separated but the separated orbits often form a large subset of V, e.g., one that is

Zariski-open and dense in V. We may take the beginning point of invariant theory to be the study of these separated orbits. There are three closely connected questions which we may state (somewhat vaguely) as follows:

(A) What is the structure of the algebra $K[V]^G$? This question starts with finite generation: are there elements $f_1,\ldots,f_m$ in $K[V]^G$ so that each element in $K[V]^G$ is a polynomial in $f_1,\ldots,f_m$? If so, the question as to whether an orbit $G\cdot v$ is separated looks easier, at least, since $\{v' \in V: f(v') = f(v)$ for all $f \in K[V]^G\} = \{v' \in V: f_i(v') = f_i(v)$ for $i = 1,\ldots,m\}$.

(B) How may the elements in $K[V]^G$ be used to define canonical forms on V? This question may be taken as a type of the same question encountered in matrix algebra or as one involving structure on the class of orbits. As an example of the latter, we might ask whether the separated orbits constitute an algebraic variety?

(C) How can invariant polynomial be constructed? In a sense, this question goes beyond the existence questions raised in Problems (A) and (B) and asks whether the generators of $K[V]^G$ can be written down explicitly or if canonical forms can be defined by certain explicitly determined invariant polynomials.

Example 1. Let $M_n(K)$ be the vector space consisting of all nxn matrices over K. The group $GL_n(K)$ acts on $M_n(K)$ via $g\cdot X = gXg^{-1}$. Given any matrix X in $M_n(K)$, we may determine its characteristic polynomial. The coefficients of the characteristic polynomial may be considered to be polynomial functions on $M_n(K)$. As such, they are invariant with respect to the action of $GL_n(K)$ and, indeed, may be shown to generate the algebra of invariant polynomials. Furthermore, these coefficients are algebraically independent, so the algebra of invariant polynomials is a polynomial algebra in n variables. The separated orbits are precisely the orbits of those X in $M_n(K)$ which have n distinct eigen-values.

<u>Example</u> 2. Let us fix a non-negative integer d and let V_d be the vector space over $\mathbb{C}$ consisting of all <u>binary forms</u> of degree d in the variables x and y. The space V_d has a basis $\binom{d}{i} x^{d-i} y^i$ for i = 0,1,...,d. The group $G = SL_2(\mathbb{C})$ acts (via multiplication on the left) on the vector space $\mathbb{C}^2$ consisting of all 2x1 column matrices. This gives an action of G on $\mathbb{C}[x,y]$. In particular, we have

$$(a_{ij})\cdot x = a_{22}x - a_{12}y \text{ and } (a_{ij})\cdot y = -a_{21}x + a_{11}y.$$

Relative to this action, we have an action of G on V_d and, so, an algebra of invariant polynomials $\mathbb{C}[V_d]^G$. Hilbert showed that the separated orbits in V_d are those of forms f in V_d which have no factor of multiplicity $\geq d/2$. He achieved this result without ever writing down the generators of $\mathbb{C}[V_d]^G$; this theme was taken up in recent times by Mumford and has developed into modern geometric invariant theory. The study of canonical forms in $\mathbb{C}[V_d]^G$ is somewhat different and we postpone our discussion of it.

The group G also acts on the product $V_d \times \mathbb{C}^2$ via $g.(f,v) = (g\cdot f, g\cdot v)$. The invariants of this action were called <u>covariants</u> in the nineteenth century. Covariants were of interest in themselves and also for their applications to the study of $\mathbb{C}[V_d]^G$ and canonical forms in V_d. For example, Gordon proved in 1868 that $\mathbb{C}[V_d]^G$ is a finitely generated algebra over $\mathbb{C}$ by using a constructive argument involving covariants called <u>transvectants</u>.

<u>Question (A): structure of</u> $K[V]^G$

We mentioned that Gordan proved that $\mathbb{C}[V_d]^G$ is finitely generated. For arbitrary groups G and rational representations V, much more is known now, mainly due to the work of Hilbert, Weyl, Schiffer, Mumford, Nagata, and Haboush. Indeed, if G is a semi-simple algebraic group, then $K[V]^G$ is always finitely generated. Furthermore, if K has characteristic 0, then Hochster and

Roberts proved that $K[V]^G$ is Cohen-Macaulay, i.e., a free module over a polynomial subalgebra. (More is known; $K[V]^G$ is actually a Gorenstein ring but not, in general, a complete intersection.)

If G is not semi-simple (or reductive), then $K[V]^G$ may or may not be finitely generated. K. Pommerening ("Invariants of unipotent groups") summarizes the current state of our knowledge, here. Interestingly enough, one critical idea in this question comes from an interpretation of covariants in the study of binary forms. (In this setting, it is due to M. Roberts and dates to 1861.) Let $U = \{(a_{ij}) \in SL_2(\mathbb{C}) : a_{21} = 0,\ a_{11} = a_{22} = 1\}$. The algebras $\mathbb{C}[V_d]^U$ and $\mathbb{C}[V_d \times \mathbb{C}^2]^G$ are isomorphic. Thus, polynomials fixed by U (called semi-invariants) were used to construct covariants and, conversely, the "leading term" of any covariant gave a semi-invariant. With this isomorphism at hand, the finite generation (and Cohen-Macaulay property) of the algebra $\mathbb{C}[V_d]^U$ now follow from the analogous facts for $SL_2(\mathbb{C})$.

In Example 1, above, we considered the action of $GL_n(K)$ on $M_n(K)$ given by $g \cdot X = gXg^{-1}$. Let $D_n(K)$ consist of all the diagonal matrices in $M_n(K)$. The permutation group on n letters acts on $D_n(K)$ in the natural way. Furthermore, the invariants of $GL_n(K)$ on $M_n(K)$ are isomorphic to the invariants of the permutation group acting on $D_n(K)$. E. Formanek ("The invariants of nxn matrices") considers the action of $GL_n(K)$ on $M_n(K) \times M_n(K) \times \ldots \times M_n(K)$ given by $g \cdot (X_1, X_2, \ldots, X_r) = (gX_1 g^{-1}, gX_2 g^{-1}, \ldots, gX_r g^{-1})$. The generators for the algebra of invariant polynomials may be given explicitly using the trace function, but now there are relations among the generators. These relations give information about nilalgebras and, more generally, algebras satisfying polynomial identities. As would be expected, permutation groups play an important role in this study.

<u>Question (B): canonical forms</u>

We mentioned earlier that for the action of $G = SL_2(\mathbb{C})$ on V_d, the separated orbits in V_d are those of forms having no linear factor of multiplicity $\geq d/2$. For such forms, a good structure (in the algebraic-geometric sense) can be placed on the class of orbits. The canonical form problem for binary forms is concerned with writing an $f \in V_d$ as $\sum_{i=1}^{m} (a_i x+b_i y)^d$. This was studied with great success by Sylvester but there are still some unresolved questions. A. Lascoux ("Forme canonique d'une forme binaire") discusses the case where d is odd and J.P.S. Kung ("Canonical forms for binary forms of even degree") the case where d is even. The information coming from the invariant polynomials in $\mathbb{C}[V_d]^G$ is not sufficient to address the questions surrounding these canonical forms. Rather, the key to unlocking the questions comes from the use of certain explicitly given covariants, such as the catalecticant, the canonizant, and certain covariants discovered by Gundelfinger.

Much the same situation arises for the action of SL(V) on the spaces $\bigwedge^k(V)$. Let us illustrate this in the case where dim V = 6 and $SL_6(C)$ acts on $\bigwedge^3(V)$. In this case, Schouten discovered four canonical forms, namely: (I) bcd, (II) bcd + bef, (III) bcd + bfg + cef, (IV) bcd + efg. (Here, the symbols b,c,d,e,f,g denote linearly independent vectors in V and the "wedge product" is denoted by juxtaposition.) The separated orbits correspond to form (IV) and there is (essentially) only one invariant polynomial. The other forms may be defined through the use of covariants, where the definition of covariant is now extended to mean the invariants of SL(V) acting on $\bigwedge^k(V) \times V \times \ldots \times V$, where V appears up to 6 times. These facts - covariants define all canonical forms on $\bigwedge^k(V)$ but invariants do not - may be shown to hold quite generally. (The extended definition of covariant given above has its roots in a classical theorem of nineteenth century invariant theory known as Gram's theorem.)

A. Nijenhuis[1]("The equivalence problem for tensor fields") considers a similar classification problem which arises in the study of differentiable manifolds. The objects to be classified are tensor fields (rather than skew -

---

[1] Added in print: This paper is not included in this volume as it is due to appear elsewhere.

symmetric tensors) and equivalence is defined with respect to diffeomorphisms (rather than SL(V)). Differential invariants play the role of invariant polynomials. Some examples of differential invariants are known but, as yet, there is no systematic procedure available for constructing them.

Question (C) : constructing invariants

Ideally, we would like (in some way) to explicitly describe some or all invariant polynomials. This is possible, for example, in Example 1 above. However, even in the case of binary forms, there are major difficulties. Mathematicians in the nineteenth century devised a symbolic technique for generating such invariants. Cayley's symbolism has its roots in the use of differential operators to produce invariants. The German symbolic technique, as developed by Aronhold, Clebsch, and Gordan, is algebraic and begins with the "symbolic" representation of an element f in V_d as $(ax + by)^d$. Eventually, it was shown that the basic invariants for binary forms are determinants symbolically. This has been extended now to arbitrary symmetric and skew-symmetric tensors.

P. Olver ("Invariant theory and differential equations") shows that several questions in differential equations can be attacked by a transform technique which changes differential operations into algebraic ones. This transform approach is closely related to the symbolic method. Furthermore, the notion of transvectant (as arising in the study of binary forms) can be generalized and becomes important in questions involving the divergence.

We noted earlier that $K[V]^G$ is always finitely generated when G is semi-simple; we now ask whether it is possible to explicitly display the generators of $K[V]^G$. In the case of binary forms, Hilbert attacked this problem in a totally unexpected way. Let $N_d = \{v \in V_d : f(v) = 0$ for every homogeneous, non-constant, polynomial $f \in \mathbb{C}[V_d]^G\}$. A form in N_d is called a <u>non-stable form</u>. Hilbert showed that there are finitely many invariant polynomials $f_1, \ldots, f_m$

so that $N_d = \{v \in V_d : f_i(v) = 0 \text{ for } i = 1,\ldots,m\}$. Furthermore, for such a set of invariant polynomials, $\mathbb{C}[V_d]^G$ is integral over $\mathbb{C}[f_1,\ldots,f_m]$. Hence, the construction of generators for $\mathbb{C}[V_d]^G$ is closely related to the geometric study of N_d.

These concepts may be extended to the actions of arbitrary semi-simple groups on finite-dimensional vector spaces. G. Kempf ("Constructing invariants") explains the basic ideas and shows how they lead to explicit (though, huge) bounds on the degrees of the generators for $K[V]^G$.

The invariants in $K[V]^G$ may be constructed, once an explicit expression for a basis of V is known, through the technique of protomorphs ("Constructing invariants via Tschirnhaus transformations"). This technique comes from the classical theory of binary forms and gives information on $K[V]^G$ and, also, algebras of invariant polynomials for certain subgroups of G.

F. G.

INVARIANTS OF UNIPOTENT GROUPS

A survey

Klaus Pommerening
Fachbereich Mathematik
der Johannes-Gutenberg-Universität
Saarstraße 21
D-6500 Mainz
Federal Republic of Germany

I'll give a survey on the known results on finite generation of invariants for nonreductive groups, and some conjectures.

You know that Hilbert's 14th problem is solved for the invariants of reductive groups; see [12] for a survey. So the general case reduces to the case of unipotent groups. But in this case there are only a few results, some negative and some positive. I assume that k is an infinite field, say the complex numbers, but in most instances an arbitrary ring would do it.

1. BASIC RESULTS

a) Nagata's counterexample (1958): Let U be a subgroup of the n-fold product G_a^n of the additive group, canonically embedded in GL_{2n}

$$U \leq \begin{pmatrix} \begin{smallmatrix} 1 & * \\ 0 & 1 \end{smallmatrix} & & \\ & \ddots & \\ & & \begin{smallmatrix} 1 & * \\ 0 & 1 \end{smallmatrix} \end{pmatrix} \leq GL_{2n},$$

such that U is given by 3 'general' linear relations. Then $k[X]^U$ is not finitely generated, where $X = (X_1, \ldots X_{2n})$, if n is a square $= r^2 \geq 16$ (at least if k contains enough transcendental elements), cf. [14]. All known counterexamples derive from this one!

Chudnovsky claims, but apparently never published a proof, that $n \geq 10$ suffices. The argument in [1] is not convincing, but there is more evidence in [2] and [15]. For the proof (with $n \geq 10$) one needs the following result:

- ▶ There is a set S of n points in the affine plane $\mathbb{A}^2$ with the property: Each nonzero polynomial $f \in k[Y_1, Y_2]$ that vanishes of order at least t in each $p \in S$ has degree $d > t \cdot \sqrt{n}$ (t any integer ≥ 1).

Let $\omega_t(S)$ be the minimum of the degrees of such polynomials; then the assertion is $\omega_t(S) > t\cdot\sqrt{n}$. Now the quotient $\omega_t(S)/t$ decreases to a limit $\Omega(S)$ when t goes to infinity; $\Omega(S)$ is called the singular degree of S. In general $\Omega(S) \leq n$. Chudnovsky's claim is:

▶ If S is generic, then $\Omega(S) = n$.

This gives $\omega_t(S) \geq t\cdot\sqrt{n}$. However, if n is not a square, we have the desired strict inequality because $\omega_t(S)$ is an integer. And in the case where n is a square, we can take Nagata's argument.

b) <u>Popov's theorem</u> (1979) is the converse of the invariant theorem for reductive groups. So for an affine algebraic group G the following statements are equivalent:

(i) G is reductive.
(ii) Whenever G acts rationally on a finitely generated algebra A, then the invariant algebra A^G is finitely generated.

See [19]. This means that for a nonreductive group there can't be a general positive answer.

c) A <u>positive result</u> goes back to <u>Zariski</u> (1954): If a group G acts on a finitely generated algebra A such that the invariant algebra A^G has transcendence degree at most 2, then A^G is finitely generated, cf. [14]. A useful geometric version is:

COROLLARY 1. If an affine algebraic group G acts on an affine variety X and there is an orbit of codimension ≤ 2, then $k[X]^G$ is finitely generated.

<u>Proof</u>. Assume (without loss of generality) that X is normal. Then

$$\operatorname{trdeg} k[X]^G \leq \dim X - \max\{\dim G\cdot x \mid x \in X\} \leq 2.$$

COROLLARY 2. If trdeg $A \leq 3$, then A^G is finitely generated.

<u>Proof</u>. Assume that G acts effectively. If G is finite, we are done. Else trdeg $A^G \leq 2$.

For linear actions we can do one more step:

COROLLARY 3. If G acts linearly on the polynomial algebra $k[X] = k[X_1,X_2,X_3,X_4]$, then $k[X]^G$ is finitely generated.

<u>Proof</u>. Assume that G acts effectively. Without changing $k[X]^G$ we may assume that G is Zariski-closed in GL_4. If G is finite, we are done. Else $\dim G \geq 1$ and G is reductive or has a 1-dimensional unipotent normal subgroup N. The algebra $A = k[X]^N$ is finitely generated by Weitzenböck's theorem (see below 2.1b), trdeg $A \leq 3$, $k[X]^G = A^G$.

d) Some other positive results derive from <u>Grosshans's principle</u> [6]: Let an algebraic group G act rationally on a k-algebra A, and H be a closed subgroup of G. Then

$$A^H \cong (k[G]^H \otimes A)^G.$$

For the <u>proof</u> let $G \times H$ act on $k[G] \otimes A$ as follows: G acts diagonally by left translation and H acts on k[G] by right translation. Then take the invariants in the two possible different ways (using an obvious isomorphism).

If G is reductive and A finitely generated, this reduces the question, whether A^H is finitely generated, to the <u>one</u> algebra $k[G]^H$ that is also the global coordinate algebra of the homogeneous space G/H.

2. APPLICATIONS OF THE GROSSHANS PRINCIPLE

For ring theoretic properties of A^H it may be useful to look at the isomorphism of 1.d. For example an unpublished result of Boutot is:

- ▶ Let char k = 0 and G reductive, acting on a finitely generated k-algebra B with only rational singularities. Then B^G also only has rational singularities; in particular B^G is Cohen-Macaulay.

The question whether $k[G]^H$ has rational singularities, seems to be rather difficult, and I don't dare making a conjecture; but there are some known examples. If that holds, and A only has rational singularities, then also $k[G]^H \otimes A$ and hence A^H only have rational singularities.

The Grosshans principle has several important special cases that were known earlier, but derived with more pains:

1.) Let $G = SL_2$ and H be the maximal unipotent subgroup consisting of upper triangular matrices. Then $k[G]^H$ is the coordinate algebra k[V] of the affine plane $V = \mathbb{A}^2$, because H is the stabilizer G_x of the point x = (1,0) whose orbit $G \cdot x = \mathbb{A}^2 - \{0\}$ is dense and isomorphic to G/H and has a boundary of codimension 2. Here are two interesting applications of this situation:

a) Let A be the coordinate algebra $k[R_d]$ of the vector space R_d of binary forms of degree d. Then we get the isomorphism

$$k[V \oplus R_d]^G \xrightarrow{\sim} k[R_d]^H$$

between 'covariants' and 'seminvariants', given by evaluating a covariant F at the point x,

$$F \longrightarrow F((1,0), -),$$

where the image is the 'Leitglied' (leading term) of the covariant. This result goes back to Roberts (around 1870).

b) Let char k = 0 and A be the coordinate algebra $k[W]$ of an arbitrary rational (finite dimensional) G_a-module W. Then the representation of G_a extends to SL_2 via the embedding by the Jordan normal form. Therefore the invariant algebra $k[W]^H \cong k[V \oplus W]^G$ is finitely generated. This is Seshadri's proof [20] of Weitzenböcks theorem (1932). Fauntleroy recently found a proof of this theorem in positive characteristic, see these conference proceedings or [5]. The proof is a skillful elaboration of the given one in characteristic 0 but, strictly speeking, doesn't depend on Grosshans's principle.

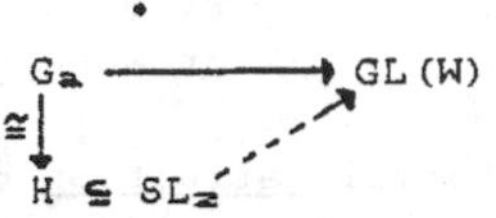

2.) Somewhat more generally we can take G reductive and H, a maximal unipotent subgroup of G. The principle for this case was observed by several people, for the first time (in characteristic 0) by Hadžiev [10], see also [6] and [21].

3.) Now let $G = GL_n$ act on the polynomial ring $k[X] = k[X_{ij} | 1 \leq i, j \leq n]$ in a matrix of indeterminates by left translation. Let H be a subgroup of SL_n such that $k[X]^H$ is finitely generated. Then for any affine algebra A on which GL_n (or a reductive group G between H and GL_n) acts rationally, the invariant algebra A^H is finitely generated. This is a qualitative version of the old principle: 'If you know the invariants of n vectors, you know all invariants.' (Capelli 1887) - The n vectors are the columns of the n-by-n matrix X.

The <u>proof</u> is two lines:

$$A^H \cong (k[GL_n]^H \otimes A)^{GL_n},$$

(I interchanged left and right translation, but that doesn't matter) and

$$k[GL_n]^H = k[X][1/\det]^H = k[X]^H[1/\det]$$

because $H \leq SL_n$ and det is SL_n-invariant.

Note that we need an action of a bigger reductive group G containing H - of course this is a disadvantage, but in view of Nagata's counterexample it even looks surprisingly good.

3. GROSSHANS SUBGROUPS

The following seems to be a good substitute of Hilbert's 14th problem:

> Find the Grosshans subgroups of GL_n
> or more generally of a reductive group G.

The formal definition of a Grosshans subgroup H of an affine algebraic group is: H is closed, G/H is quasiaffine, k[G/H] is finitely generated. The technical condition 'G/H quasiaffine' is automatic if H is unipotent. Let me give 3 examples:

a) By Hadžievs result the maximal unipotent subgroups are Grosshans, even if G is not reductive.

b) The existence of non-Grosshans subgroups follows from Nagata's counterexample: There must be a situation

$$GLn \geq U \geq V, \quad U \text{ and } V \text{ unipotent with } \dim U/V = 1,$$

such that U is Grosshans and V is not.

c) Generic stabilizers often are Grosshans subgroups. The following theorem generalizes a result by Grosshans [7]. Since some people recently were interested in it, I give the proof here. I would be happy to see a substantial application.

THEOREM. Let X be a factorial affine variety and G, an affine algebraic group acting on X. Then X has a dense open subset U such that the stabilizer G_x is a Grosshans subgroup of G for all $x \in U$.

Remark. Instead of 'factorial' the following condition suffices: X is normal and each G-invariant divisor on X has finite order in the divisor class group Cl(X), cf. [18].

Proof. I may assume G connected. There is a function $f \in k[X]$ such that the principal open subset X_f is G-stable and $k(X)^G$ is the quotient field of $k[X_f]^G$; this is well-known, cf. [13]. Choose functions $f_1, \ldots, f_n \in k[X_f]^G$ that generate the field $k(X)^G$. Let R be the algebra generated by $f_1, \ldots, f_n$ and Y be an affine model of R. Then $k(Y) = k(X)^G$, and the induced morphism $\pi: X_f \longrightarrow Y$ is dominant.

Now let $m = \max\{\dim G\cdot x \mid x \in X\}$ be the maximal orbit dimension. The set $Z = \{x \in X_f \mid \dim G\cdot x = m\}$ is G-stable and open dense in X, and $\dim Y = \dim X - m$. Since X_f is factorial, $X_f - Z = V(h) \cup A$ for a function $h \in k[X_f]$ with $\dim A \leq \dim X - 2$. Clearly X_{fh} is G-stable. Restricting π gives a dominant morphism $\sigma: X_{fh} \longrightarrow Y$. The fibers of σ are G-stable, and

$$\sigma^{-1}\sigma x = (\sigma^{-1}\sigma x \cap Z) \cup (\sigma^{-1}\sigma x \cap A) \quad \text{for all } x \in Z.$$

There is a dense open part $W \subseteq Z$ such that $\sigma^{-1}y$ has pure dimension m for all $y \in W$. Shrinking W we may assume that

$$\dim(\sigma^{-1}y \cap A) \leq m-2 \quad \text{for all } y \in W.$$

Now $U = Z \cap \sigma^{-1}W$ is G-stable and open dense in X. Let $x \in U$. Then the closure $\overline{G\cdot x}$ in X_{rn} is an irreducible component of $\sigma^{-1}\sigma x$ – compare the dimensions. In $\overline{G\cdot x} - G\cdot x$ there is no $z \in Z$ since there is no room for the m-dimensional orbit $G\cdot z$. Thus $\overline{G\cdot x} - G\cdot x \subseteq \sigma^{-1}\sigma x \cap A$, and

$$\dim(\overline{G\cdot x} - G\cdot x) \leq \dim(\sigma^{-1}\sigma x \cap A) \leq m - 2 \leq \dim G\cdot x - 2.$$

Therefore G_n is a Grosshans subgroup of G. ∎

There are some natural conjectures:

<u>Conjecture m</u>: Each m-dimensional unipotent subgroup is Grosshans.

This conjecture is false when $m = r^2 - 3$ and $r \geq 4$, and probably false when $m \geq 7$. Conjecture 1 is true by Weitzenböck´s theorem, and I guess this is the only positive case! Since nobody seems to have an approach to this problem, I make another conjecture:

<u>Conjecture A</u>: Each regular unipotent subgroup of a reductive group is Grosshans.

´Regular´ means ´normalized by a maximal torus´, or, more concretely, given by a closed subset of the root system. For GL_n it means that the subgroup is defined by relations of the type $X_{ij} = 0$. The following example shows what this means:

$$\boxed{\begin{matrix} 1 & 0 & * & 0 \\ & 1 & 0 & * \\ & & 1 & 0 \\ & & & 1 \end{matrix}} = \left\{ \begin{bmatrix} 1 & 0 & a & 0 \\ & 1 & 0 & b \\ & & 1 & 0 \\ & & & 1 \end{bmatrix} \;\middle|\; a, b \in k \text{ arbitrary} \right\}.$$

Such a pattern of zeroes and stars above the diagonal gives a subgroup of GL_n, if and only if it is the incidence matrix of a strict ordering of the set $\{1,\ldots,n\}$.

Conjecture A is true for the unipotent radicals $H = R_u(P)$ of the parabolic subgroups P. This was shown by Hochschild and Mostow 1973 (for characteristic 0) [11], and by Grosshans 1983 (for the general case) [8]. Grosshans recently extended this result in several ways [9].

4. INVARIANT MINORS

My own contribution in [16], [17] is a large class of examples for GL_n - but unfortunately I have no general proof of conjecture A, not even for GL_n. My approach is the explicit determination of $k[X]^H$, where $X = (X_{ij})$ is the n-by-n matrix of indeterminates and $H \leq GL_n$ regular unipotent. It looks promising because a lot of invariants are obvious: Consider a minor

$$\begin{vmatrix} X_{i_1j_1} & \cdots & X_{i_1j_m} \\ \vdots & & \vdots \\ X_{i_mj_1} & \cdots & X_{i_mj_m} \end{vmatrix},$$

shortly represented by the row $(i_1 \ldots i_m | j_1 \ldots j_m)$. When is it invariant? For the group at the end of section 3 we have

$$\begin{bmatrix} 1 & 0 & a & 0 \\ & 1 & 0 & b \\ & & 1 & 0 \\ & & & 1 \end{bmatrix} \cdot \begin{bmatrix} X_{11} & \ldots \\ X_{21} & \ldots \\ X_{31} & \ldots \\ X_{41} & \ldots \end{bmatrix} = \begin{bmatrix} X_{11}+aX_{31} & \ldots \\ X_{21}+bX_{41} & \ldots \\ X_{31} & \ldots \\ X_{41} & \ldots \end{bmatrix}.$$

So our minor is invariant if and only if the following is true:

▶ If it contains the row index 1, then it also contains 3,
if it contains the row index 2, then it also contains 4.

This is because then H acts by elementary row operations. In general we read the condition off the constellation of stars:

▶ If the box representing the group contains stars at positions l_1, ..., l_m in the i-th row, then the minor has to contain the row indices l_1, ..., l_m along with i.

So we know the invariant minors. I would like to prove:

Conjecture B: The invariant algebra $k[X]^H$ is generated by the (finitely many) invariant minors.

This would imply Conjecture A for GL_n. In fact I can prove a much stronger result, but only for a large class of unipotent subgroups that however contains the unipotent radicals of the parabolics as simplest special cases. Since the proof (and even the statement of the result) uses some complicated combinatorial methods, I'll give only a very simple example that, of course, is not new.

Take n = 2 and H, the maximal unipotent subgroup consisting of upper triangular matrices. The invariant minors are:

$$(1\ 2|1\ 2) = \det, \quad (2|1) = X_{21}, \quad (2|2) = X_{22},$$

because, if we have the row index 1 we also must have the row index 2, so the minor is the full determinant. Now $k[X][1/X_{22}] = R[1/X_{22}][X_{12}]$, where R is the algebra generated by the invariant minors; H acts trivially on $R[1/X_{22}]$ and maps X_{12} to $X_{12} + sX_{22}$ with $s \in k$ arbitrary. Therefore

$$k[X][1/X_{22}]^{H} = R[1/X22].$$

This kind of argument goes through for general n: There is a product ϵ of invariant minors such that $k[X][1/\epsilon]^{H} = R[1/\epsilon]$. This means that the analogue of Conjecture B for rational functions is true. However, as it is often the case, it is a major problem to get rid of this denominator.

Let me continue the example. We have

$$k[X]^{H} = k[X] \cap R[1/X_{22}] \supseteq R.$$

To get equality I have to show: If $f \in k[X]$ and $X_{22}f \in R$, then $f \in R$ - by induction I may assume $r = 1$. This is the hard core of the proof, that however is no problem for this example. Write a product of minors in the form of a bitableau, say

$$\left[\begin{array}{cc|cc} 1 & 2 & 1 & 2 \\ 1 & 2 & 1 & 2 \\ 1 & & 1 & \\ 1 & & 2 & \\ 2 & & 1 & \end{array}\right].$$

These bitableaux span $k[X]$. When the columns increase, we have a 'standard' bitableau. The given one is not standard because of its last entry. The straightening law by Rota and others, see [3] for example, says (for the general case of n-by-n matrices):

▶ The standard bitableaux are a basis of $k[X]$.

This generalizes some classical determinant identities. In the example we have

$$\left(\begin{array}{c|c} 1 & 2 \\ 2 & 1 \end{array}\right) = \left(\begin{array}{c|c} 1 & 1 \\ 2 & 2 \end{array}\right) - (1\ 2 | 1\ 2);$$

in a similar way each bitableau obviously is a linear combination of standard ones.

Now call a bitableau 'admissible', if each of its rows represents an invariant minor. Then R is spanned by the admissible bitableaux. In our example the admissible standard bitableaux are a basis of R: 'Admissible' means that there is no row $(1|\ldots)$; but then the bitableau is already standard.

Now take $f \in k[X]$ such that $X_{22}f \in R$, and write it as a linear combination of standard bitableaux:

$$f = c_1T_1 + \ldots + c_rT_r \quad \text{(with nonzero coefficients } c_i\text{)}.$$

$$X_{22}f = \sum c_iT_iX_{22} \in R$$

is a linear combination of standard bitableaux T_iX_{22}. These have to be admissible, so have their parts T_i. Therefore $f \in R$.

In the general case, under suitable conditions on H, the proof goes the same way except that the minors of an n-by-n matrix behave a lot more complicated. This is where the nontrivial combinatorial techniques come in.

Finally let me note that conjecture B is true in the cases

a) $\dim H \leq 3$,
b) $n \leq 4$.

The first examples where the Grosshans property is unknown are

$$\begin{pmatrix} 1 & 0 & * & * & 0 \\ & 1 & 0 & * & * \\ & & 1 & 0 & 0 \\ & & & 1 & 0 \\ & & & & 1 \end{pmatrix} \quad \text{and} \quad \begin{pmatrix} 1 & 0 & 0 & * & 0 \\ & 1 & 0 & * & * \\ & & 1 & 0 & * \\ & & & 1 & 0 \\ & & & & 1 \end{pmatrix},$$

and this are the only exceptions for n = 5.

REFERENCES

[1] G.V.CHOODNOVSKY: Sur la construction de Rees et Nagata pour le 14• problème de Hilbert. C.R.Acad.Sci.Paris 286 (1978), A1133-1135.

[2] G.V.CHUDNOVSKY: Singular points on complex hypersurfaces and multidimensional Schwarz lemma. Séminaire Delange-Pisot-Poitou 1979-80, Birkhäuser, Prog.Math. 12 (1981), 29-69.

[3] J.DÉSARMÉNIEN, J.P.S.KUNG, G.-C.ROTA: Invariant theory, Young bitableaux, and combinatorics. Adv.Math. 27 (1978), 63-92.

[4] A.FAUNTLEROY: Algebraic and algebro-geometric interpretations of Weitzenbock's problem. J.Algebra 62 (1980), 21-38.

[5] - : Factorial algebraic schemes defined by reflexive modules. Preprint 1985. (cf. the talk at this conference).

[6] F.GROSSHANS: Observable groups and Hilbert's fourteenth problem. Amer.J.Math. 95 (1973), 229-253.

[7] - : Open sets of points with good stabilizers. Bull.Amer.Math. Soc. 80 (1974), 518-521.

[8] - : The invariants of unipotent radicals of parabolic subgroups. Invent.Math. 73 (1983), 1-9.

[9] - : Hilbert's fourteenth problem for non-reductive groups. Preprint 1985.

[10] DŽ.HADŽIEV: Some questions in the theory of vector invariants. Math.USSR Sbornik 1 (1967), 383-396.

[11] G.HOCHSCHILD, G.D.MOSTOW: Unipotent groups in invariant theory. Proc.Nat.Acad.Sci.USA 70 (1973), 646-648.

[12] J.E.HUMPHREYS: Hilbert's fourteenth problem. Amer.Math.Monthly 85 (1978), 341-353.

[13] D.LUNA: Slices étales. Bull.Soc.Math.France Mém. 33 (1973), 81-102.

[14] M.NAGATA: Lectures on the Fourteenth Problem of Hilbert. Tata Institute, Bombay 1965.

[15] P.PHILIPPON: Interpolation dans les espaces affines. Séminaire Delange-Pisot-Poitou 1980-81. Birkhäuser, Prog.Math. 22 (1982), 221-235.

[16] K.POMMERENING: Invarianten unipotenter Gruppen. Math.Z. 176 (1981), 359-374.

[17] - : Ordered sets with the standardizing property and straightening laws for algebras of invariants. Adv. Math. (to appear).

[18] V.L.POPOV: On the stability of the action of an algebraic group on an algebraic variety. Math.USSR Izv. 6 (1972), 367-379.

[19] - : Hilbert's theorem on invariants. Soviet Math. Dokl. 20 (1979), 1318-1322.

[20] C.S.SESHADRI: On a theorem of Weitzenböck in invariant theory. J.Math.Kyoto Univ. 1 (1962), 403-409.

[21] Th.VUST: Sur la théorie des invariants des groupes classiques. Ann.Inst.Fourier 26 (1976), 1-31.

THE INVARIANTS OF $n\times n$ MATRICES

Edward Formanek*
The Pennsylvania State University
University Park, PA 16802

Table of Contents

*Partially supported by the National Science Foundation.

Introduction.

Let K be a field of characteristic zero and let U be a finite-dimensional vector ace over K with basis $u_1,\ldots,u_n$. Let

$$K[u_1,\ldots,u_n] = K[U] = K \oplus U \oplus S^2(U) \oplus S^3(U) \oplus \ldots$$

note the symmetric algebra of U over K, a polynomial ring in n variables over K. re $S^r(U)$ denotes the r-th symmetric power of U, the K-vector subspace of $K[U]$ anned by the monomials in $u_1,\ldots,u_n$ of degree r. The symmetric algebra of U is a aded K-algebra and $GL(U) = GL(n,K)$, the group of K-automorphisms of U, can be iden- fied with the group of homogeneous automorphisms of $K[U]$. If G is a subgroup of .(U), then G acts on $K[U]$ and

$$K[U]^G = K \oplus U^G \oplus S^2(U)^G \oplus \ldots$$

notes the ring of invariants of G acting on $K[U]$. The ring of invariants inherits e grading of $K[U]$.

This article is concerned with a particular ring of invariants, the ring of variants of a set of $n\times n$ matrices over K. Let U be $M_n(K)^r$, the direct sum of r pies of $M_n(K)$, with the adjoint action of $GL(n,K)$:

$$(X_1,\ldots,X_r) \longmapsto (PX_1P^{-1},\ldots,PX_rP^{-1}),$$

ere $X_1,\ldots,X_r \in M_n(K)$, $P \in GL(n,K)$. Set

$$C(n,r) = K[M_n(K)^r]^{GL(n,K)},$$

ich is called the ring of invariants of r $n\times n$ matrices. Sometimes we take a countably finite set of $n\times n$ matrices and then $C(n)$ denotes the ring of invariants.

It is useful to introduce generic $n\times n$ matrices $U_1,U_2,\ldots,$ where $U_k = (u_{ij}(k))$ $\leq i,j \leq n)$, and the $u_{ij}(k)$ are independent commuting indeterminates over K. Then

$$C(n,r) = K[u_{ij}(k) \mid 1 \leq k \leq r]^{GL(n,K)},$$

$$C(n) = K[u_{ij}(k) \mid k = 1,2,\ldots]^{GL(n,K)}.$$

The object of this article is to survey what is known about $C(n,r)$ and $C(n)$, as ell as some applications to noncommutative ring theory. We are concerned with the imultaneous invariants of $n\times n$ matrices $(r \geq 2)$ since the invariant theory of a ingle $n\times n$ matrix $(r = 1)$ is completely understood.

Theorem 1 [Sp, p. 10]. $C(n,1)$ *is a polynomial ring in* n *variables over* K. *The coefficients of the characteristic polynomial of the generic matrix* U_1 *are an independent generating set for* $C(n,1)$. *The traces of* $U_1, U_1^2, \ldots, U_1^n$ *are also an independent generating set.*

2. The First and Second Fundamental Theorems for Matrix Invariants.

Let $T(X)$ denote the trace of an $n\times n$ matrix X. If $X_1, \ldots, X_r$ are (not necessarily distinct) $n\times n$ matrices, then for any $P \in GL(n,K)$,

$$T(X_1 \ldots X_r) = T((PX_1P^{-1})\ldots(PX_r(P^{-1})).$$

Hence $T(X_1 \ldots X_r)$ is an invariant. Conversely

Theorem 2 (First Fundamental Theorem of Matrix Invariants [G, Theorem 16.2], [P2, Theorem 1.3], [S, Theorem 1]. $C(n)$ *is generated as a* K-*algebra by the traces* $T(U_{i_1}\ldots U_{i_t})$, *where* $U_{i_1}\ldots U_{i_t}$ *is a monomial in the generic matrices* $U_1, U_2, \ldots$.

The second fundamental theorem gives all multilinear relations among the traces. In order to state it, we have to introduce some terminology. We assume some familiarity with the representation theory of the symmetric group, for which [J-K] is a good reference. For us, the single most important aspect of the representation theory of the symmetric group on r letters is the one-to-one correspondence between simple factors of the group algebra and Young diagrams of size r (or partitions of r).

Let σ be an element of S_r, the symmetric group of permutations of $\{1,\ldots,r\}$, and write σ as a product of disjoint cycles

$$\sigma = (a_1 \ldots a_{k_1})(b_1 \ldots b_{k_2})(c_1 \ldots c_{k_2}) \ldots ,$$

where 1-cycles are included, so that each of the digits $1,\ldots,r$ occurs exactly once. Define the *associated trace function* $T : M_n(K)^r \to K$ by

$$T_\sigma(X_1,\ldots,X_r) = T(X_{a_1}\ldots X_{a_{k_1}})T(X_{b_1}\ldots X_{b_{k_2}})T(X_{c_1}\ldots X_{c_{k_3}})\ldots ,$$

where T is the usual trace. Note that if $U_1,\ldots,U_r$ are the generic $n\times n$ matrices, then $T_\sigma(U_1,\ldots,U_r)$ lies in $C(n)$.

Theorem 3 (Second Fundamental Theorem of Matrix Invariants [P2, Theorem 4.3], [R, Proposition 1]). *Let* KS_r *be the group algebra of* S_r *over* K, *and let* $J(n,r)$ *be the two-sided ideal of* KS_r *which is the sum of all simple factors of* KS_r *corresponding to Young diagrams with* $\geq n+1$ *rows. Define a* K-*linear map* $\varphi: KS_r \to C(n)$ *by*

$$\varphi(\Sigma a_\sigma \sigma) = \Sigma a_\sigma T_\sigma(U_1,\ldots,U_r).$$

Then $\operatorname{Ker}\varphi = J(n,r)$.

There are two other characterizations of the ideal $J(n,r)$:

(I) If V is a vector space of dimension n and S_r acts on $V^{\otimes r}$ by place permutation

$$\sigma(v_1 \otimes \ldots \otimes v_r) = v_{\sigma^{-1}(1)} \otimes \ldots \otimes v_{\sigma^{-1}(r)},$$

then $J(n,r)$ is the kernel of the action of KS_r on $V^{\otimes r}$.

(II) $J(n,r) = 0$, if $r \leq n$; and if $r \geq n+1$, $J(n,r)$ is the two-sided ideal of KS_r generated by

$$\Sigma\{(\operatorname{sign}\sigma)\sigma \mid \sigma \in S_{n+1}\}.$$

In a sense which will be made precise later in Theorem 18, all relations among traces are consequences of the fundamental trace identity for $n\times n$ matrices: If $X_1,\ldots,X_{n+1} \in M_n(K)$, then

$$\sum_{\sigma\in S_{n+1}} (\operatorname{sign}\sigma)T_\sigma(X_1,\ldots,X_{n+1}) = 0.$$

The fundamental trace identity is actually the multilinearization of the Cayley-Hamilton polynomial, as we now illustrate for $n = 2$.

Start with the Cayley-Hamilton Theorem for 2×2 matrices:

(1) $$X^2 - T(X)X + \operatorname{Det}(X) = 0.$$

Use Newton's formulas for the elementary symmetric functions as polynomials in the power symmetric functions to experss the determinant $\operatorname{Det}(X)$ in terms of traces of powers of X.

(2) $$X^2 - T(X)X + \tfrac{1}{2}[T(X)^2 - T(X^2)] = 0.$$

Multilinearize

(3) $$X_1X_2 + X_2X_1 - T(X_1)X_2 - T(X_2)X_1 + T(X_1)T(X_2) - T(X_1X_2) = 0.$$

Multiply on the right by X_3 and take the trace

$$T(X_1X_2X_3) + T(X_2X_1X_3) - T(X_1)T(X_2X_3) - T(X_2)T(X_1X_3) + T(X_1)T(X_2)T(X_3) - T(X_1X_2)T(X_3) \tag{4}$$

$$= \sum_{\sigma\in S_3} (\text{sign }\sigma) T_\sigma(X_1,X_2,X_3) = 0.$$

The implications (1) ⇒ (2) ⇒ (3) ⇒ (4) are easily seen to be reversible ((4) ⇒ (3) by the nondegeneracy of the trace), so the fundamental trace identity may be regarded as an alternate form of the Cayley-Hamilton Theorem.

Both of the fundamental theorems are translations of classical theorems on vector invariants using the $GL(n,K)$-isomorphism

$$[M_n(K)^r] \cong (V \otimes V^*)^r \cong \mathrm{Hom}_K(V^r,V^r),$$

where V is the standard $GL(n,K)$-module. This is clearly seen in the presentations of Procesi [P2], [P4].

3. General Properties of the Ring of Matrix Invaraints.

In the adjoint action $(X \mapsto PXP^{-1})$ of $GL(n,K)$ on $M_n(K)$, the center, K^*, of $GL(n,K)$ acts trivially, so it is really an action of $PGL(n,K)$. Moreover, $K^*\cdot SL(n,K)$ is Zariski dense in $GL(n,K)$ (they are equal if K is algebraically closed). Hence

$$C(n,r) = K[M_n(K)^r]^{GL(n,K)} = K[M_n(K)^r]^{SL(n,K)}.$$

A lot of information about $C(n,r)$ can be obtained from general theorems about the ring of invariants of a reductive group acting homogeneously on a polynomial ring.

Theorem 4. (1) $C(n,r)$ is a finitely generated K-algebra and hence is Noetherian.

(2) $C(n,r)$ is a Cohen-Macaulay ring.

(3) $C(n,r)$ is a unique factorization domain.

(4) $C(n,r)$ is a Gorenstein ring.

(5) If $r \geq 2$, $C(n,r)$ has transcendence degree $n^2(r-1) + 1$ over K.

(6) If $r \geq 2$, $C(n,r)$ has Krull dimension $n^2(r-1) + 1$.

Part (1) goes back to Hilbert (see [M, pp. 432-433]). Part (2) follows from a fundamental theorem of Hochster and Roberts [H-R, Main Theorem]. Parts (3) and (4) are also observations of Hochster and Roberts [H-R, Corollary 1.9]. Part (3) is based on the fact that $SL(n,K)$ has no nontrivial characters, while part (4) follows from (3) using a theorem of Murthy. Part (5) follows from an application of a theorem of Rosenlicht [Sp, p. 39] which says that

$$\text{trans.deg. } K[V]^G = \dim V - \dim G$$

if G is an algebraic subgroup of $GL(V)$ having no nontrivial characters and the set of points of V whose G-stabilizer is trivial contains a dense open subset of V. If $V = M_n(K)^r$ and $G = SL(n,K)$, the set of r-tuples $(X_1,\ldots,X_r)$ which generate $M_n(K)$ as a K-algebra contains a dense open subset of $M_n(K)^r$. Then the theorem gives

$$\text{trans.deg. } C(n,r) = rn^2 - (n^2-1) = (r-1)n^2 + 1.$$

Finally, part (6) follows from (5) since the Krull dimension equals the transcendence degree for finitely generated domains over K [Ho, p. 164].

The Cohen-Macaulay and Gorenstein properties and the Krull dimension give information about presenting $C(n,r)$ as a K-algebra. The Cohen-Macaulay property says that $C(n,r)$ is a free module over a polynomial subring, and the number of generators of the polynomial subring is the Krull dimension. Both the generators of the polynomial subring and the module generators of $C(n,r)$ over it may be taken to be homogeneous. The Gorenstein property implies certain symmetries among the degrees of the generators of $C(n,r)$.

In the next two sections we will give further clues toward finding presentations of $C(n,r)$, as well as the few cases where presentations are known.

4. Hilbert Series and Functional Equations.

As already noted, $C(n,r)$ is a graded ring. It also has a multigrading by $\mathbb{N}^r$ induced by giving the generic matrix U_k degree $(0,\ldots,0,1,0,\ldots,0)$, where the k-th coordinate is 1. More precisely, a generating trace $T(U_{i_1}\ldots U_{i_t})$ in $C(n,r)$ has degree $(\alpha_1,\ldots,\alpha_r)$, where α_k is the number of occurences of U_k in the monomial $U_{i_1}\ldots U_{i_t}$. The total degree of $\alpha = (\alpha_1,\ldots,\alpha_r) \in \mathbb{N}^r$ is defined to be $|\alpha| = \alpha_1 + \ldots + \alpha_r$. It gives the ordinary grading of $C(n,r)$.

For each $\alpha \in \mathbb{N}^r$, let $C(n,r)_\alpha$ $(C(n,r)_k)$ denote the K-vector subspace of $C(n,r)$ spanned by all homogeneous element of degree α (total degree k). Each $C(n,r)_\alpha$ and $C(n,r)_k$ is finite-dimensional over K, and $C(n,r)_k = \Sigma\{C(n,r)_\alpha \mid |\alpha| = k\}$.

Associated to $C(n,r)$ is a Hilbert series

$$H[C(n,r)] = H[C(n,r)](t_1,\ldots,t_r) = \sum_{\alpha\in\mathbb{N}^r} [C(n,r)_\alpha : K]t^\alpha,$$

where $t^\alpha = t_1^{\alpha_1}\ldots t_r^{\alpha_r}$ for $\alpha = (\alpha_1,\ldots,\alpha_r) \in \mathbb{N}^r$. The Hilbert series is an element of the ring of formal power series $\mathbb{Z}[[t_1,\ldots,t_r]]$. A one-variable Hilbert series based on total degree can likewise be defined.

There is an action of $GL(n,K)$ as a group of automorphisms of $C(n,r)$ induced by the linear action of $GL(r,K)$ on the K-vector space spanned by the generic matrices $U_1,\ldots,U_r$. The Hilbert series encodes the structure of $C(n,r)$ as a $GL(r,K)$-module as follows.

Note that each $C(n,r)_k$ is a $GL(r,K)$-module. In other words, the action of GL(r,K is homogeneous with respect to total degree. Moreover

$$H[C(n,r)_k] = \Sigma\{[C(n,r)_\alpha : K]t^\alpha \mid |\alpha| = k\}$$

is a symmetric function of $t_1,\ldots,t_r$ which is homogeneous of degree k. Recall the representation theory of $GL(r,K)$: There is an isomorphism of graded rings

$$\chi : \mathrm{Mod}[GL(r,K)] \to \mathbb{Z}[t_1,\ldots,t_r]^{S_r}$$

between the Grothendieck ring of finite-dimensional polynomial $GL(r,K)$-modules and symmetric functions of $t_1,\ldots,t_r$. It turns out that

$$\chi[C(n,r)_k] = H[C(n,r)_k]$$

In other words, the Hilbert series of $C(n,r)$ is the character of $C(n,r)$ as a $GL(r,K)$-module. For more details, see [F2, §2-6] or [F3, §3].

There are various formulas for the Hilbert series of $C(n,r)$, which can be obtained using classical methods of Molien, Schur, and Weyl. It can be expressed as a formal power series of Schur functions [F2, Theorem 12], as an integral over the unitary group $U(n,\mathbb{C})$ [F2, p.204], or, using a device of Weyl [W, p.198], as a multiple integral over the unit circle [T1, Theorem 2.1]. In theory such integrals can be evaluated by partial fractions and the calculus of residues, but in practice it usually is too difficult. In most instances where the Hilbert series of $C(n,r)$ is known, it is a byproduct of a presentation of $C(n,r)$.

In the next section we will give presentations for a few $C(n,r)$. While a presentation of a graded k-algebra yields its Hilbert series, the converse is not true. Nevertheless, the Hilbert series often can be of use in finding a presentation.

The Cohen-Macaulay property implies that $H[C(n,r)]$ is a rational function of $t_1,\ldots,t_r$. The Gorenstein property in conjunction with a theorem of Stanley [St, Theorem 6.1(i)] implies that $H[C(n,r)]$ satisfies a functional equation.

<u>Theorem 5</u> [F4, Theorem 13], [T1, Proposition 2.3]. <u>For</u> $r \geq 2$, $H[C(n,r)]$ <u>satisfies the functional equation</u>

$$H[C(n,r)](t_1^{-1},\ldots,t_r^{-1}) = (-1)^d (t_1\cdots t_r)^{n^2} H[C(n,r)](t_1,\ldots,t_r),$$

here $d = (r-1)n^2 + 1$, the Krull dimension of $C(n,r)$.

The theorem was proved for $r \geq n^2 - n$ by Formanek and in full generality by eranishi. Stanley's theorem implies that $H[C(n,r)]$ satisfies a functional equation, ut does not give its exact form.

The fact that n^2 is the exponent in the functional equation suggests that there ay be canonical presentations for $C(n,r)$ but as of now these are not enough to suggest hat form such presentations might have.

. Generators for the Ring of Matrix Invariants.

The following are all the cases $(n,r \geq 2)$ for which either a presentation or the ilbert series of $C(n,r)$ is known.

$C(2,2)$ is a polynomial ring freely generated by the traces of U_1, U_1^2, U_2, U_2^2, U_1U_2 P1]. Therefore its Hilbert series is

$$[(1-t_1)(1-t_1^2)(1-t_2)(1-t_2^2)(1-t_1t_2)]^{-1}.$$

$C(2,3)$ is a free module of rank 2 with generators 1, $T(U_1U_2U_3)$ over the polynomial ing freely generated by the traces of U_1, U_1^2, U_2, U_2^2, U_3, U_3^2, U_1U_2, U_1U_3, U_2U_3 [F2, heorem 22]. Its Hilbert series is

$$(1+t_1t_2t_3)\{[\prod_{i=1}^{3} (1-t_i)(1-t_i^2))](1-t_1t_2)(1-t_1t_3)(1-t_2t_3)\}^{-1}.$$

The Hilbert series of $C(2,r)$ has an expression as a formal power series of Schur unctions which is independent of r [F2, Theorem 21]. For $r = 4$, the Hilbert series s

$$\frac{1 + t_it_jt_k - t_1t_2t_3\Sigma t_i - (t_1t_2t_3t_4)^2}{\Pi(1-t_i)\ \Pi\ (1-t_i^2)\ \Pi\ (1-t_it_j)},$$

here the sums and products are over $1 \leq i \leq 4$, $1 \leq i < j \leq 4$, $1 \leq i < j < k \leq 4$, espectively.

For $r \leq 4$, the Hilbert series of $C(2,r)$ is obtained from that of $C(2,4)$ by etting $t_i = 0$ for $i > r$. For $r > 4$, the Hilbert series of $C(2,r)$ can be obtained rom that of $C(2,4)$ by repeated "polarizations" [F4, §3] based on the fact that, as ormal power series of Schur functions, $C(2,r)$ involves only Schur functions orresponding to partitions with ≤ 4 parts.

$C(3,2)$ is a free module of rank 2 with generators 1, $T(U_1U_2U_1^2U_2^2)$ over the olynomial ring freely generated by the traces of U_1, U_1^2, U_1^3, U_2, U_2^2, U_2^3, U_1U_2, $U_1^2U_2$, $U_1U_2^2$, $U_1U_2U_1U_2$ [T2, §3]. Its Hilbert series is

$$\frac{1 + t_1^3 t_2^3}{(1-t_1)(1-t_1^2)(1-t_1^3)(1-t_2)(1-t_2^2)(1-t_2^3)(1-t_1t_2)(1-t_1^2t_2)(1-t_1t_2^2)(1-t_1^2t_2^2)} .$$

$C(4,2)$ is a free module of rank 48 over the polynomial ring freely generated by the traces of $U_1^iU_2^j$ $(i,j \geq 0,\ 0 < i + j \leq 4)$, $U_1U_2U_1U_2$, $U_1U_2^2U_1U_2^2$, $U_1^2U_2U_1^2U_2$ [T2, §4]. Its Hilbert series is

$$f(t_1,t_2)\ [\ \Pi\ (1-t_1^it_2^j)\,|\,i,j \geq 0,\ 0 < i + j \leq 4]\ (1-t_1^2t_2^2)(1-t_1^4t_2^2)(1-t_1^2t_2^4)\}^{-1} ,$$

where $f(t_1,t_2)$ is an explicit polynomial with positive integer coefficients.

Instead of asking for explicit generators one can ask for the minimal degree of a generating set. It follows from general arguments (essentially because n^2 is the dimension of $M_n(K)$ over K) that if $C(n,n^2)$ is generated by its elements of degree $\leq d$, then so is $C(n,r)$ for any r.

The second fundamental theorem allows this question to be translated into a combinatorial problem about the group ring of the symmetric group. The point is that $C(n)$ is generated by traces of degree $\leq d$ if and only if $T(U_1 \ldots U_{d+1})$ can be expressed in terms of traces of shorter monomials. Under the homomorphism $\varphi\colon KS_{d+1} \to C(n)$ of the second fundamental theorem

$\varphi(\sigma) = T(U_{i_1} \ldots U_{i_{d+1}})$ if σ is a $(d+1)$-cycle

$\varphi(\sigma)$ = product of $k > 2$ shorter traces if
σ is the product of $k \geq 2$ disjoint cycles.

This leads to

<u>Theorem 6</u> [F5, Lemma 2]. *Let $B(d+1)$ be the K-subspace of KS_{d+1} spanned by all group elements except the $(d+1)$-cycles, and let $J(n,d+1)$ be the two-sided ideal of KS_{d+1} generated by all simple factors corresponding to Young diagrams with $> n + 1$ rows. Then $C(n)$ is generated by its elements of degree $\leq d$ if and only if $KS_{d+1} = B(d+1) + J(n,d+1)$.*

The condition on the group algebra in Theorem 6 will reappear later in connection with the index of nilpotence of a nil algebra (Theorem 20). The first part of the next theorem was originally deduced by Procesi as a corollary to a result of Razmyslov on nil algebras. A more direct proof can be obtained using Theorem 6. The second part follows from a result of Kuzmin [Ku] using Theorems 6 and 20.

Theorem 7. (1) [F5, Theorem 6]. See also [P2, Theorem 3.3], [P3, Theorem 4.3], [R, final remark]. $C(n)$ is generated by its elements of degree $\leq n^2$. (2) [Ku]. $C(n)$ is not generated by its elements of degree $< n(n+1)/2$.

The minimal degree $d(n)$ for a generating set for $C(n)$ is known only for $n \leq 3$. Trivially, $d(1) = 1$. Dubnov [D] showed that $d(2) = 3$ and $d(3) = 6$.

5. The Mixed Discriminant and the Discriminant.

By the first fundamental theorem every invariant of $n\times n$ matrices can be expressed in terms of traces. Let us illustrate the theorem by giving explicit trace formulas for two basic invariants, the mixed discriminant and the discriminant.

The mixed discriminant is the multilinear function of $n\times n$ matrices $X_1,\ldots,X_n$ defined as follows: For each permutation $\pi \in S_n$. Let $D(\pi)$ be the determinant of the $n\times n$ matrix whose i-th row is the i-th row of $X_{\pi(i)}$. The mixed discriminant is then defined by

$$D(X_1,\ldots,X_n) = \sum_{\pi\in S_n} D(\pi) .$$

The mixed discriminant satisfies

$$D(AX_1B,\ldots,AX_nB) = (\det A)(\det B)\, D(X_1,\ldots,X_n),$$

so it is an invariant. Moreover, it vanishes if $X_1,\ldots,X_n \in M_{n-1}(K)$. Hence the second fundamental theorem implies that $D(X_1,\ldots,X_n)$ is a scalar multiple of $\Sigma\{(\text{sign } \sigma)T_\sigma(X_1,\ldots,X_n) \mid \sigma\in S_n\}$. The specialization $X_i = e_{ii}$, where the e_{ij} are the standard matrix units, shows that the scalar is 1.

Theorem 8. As a function of traces, the mixed discriminant is given by

$$D(X_1,\ldots,X_n) = \sum_{\sigma\in S_n} (\text{sign } \sigma)T_\sigma(X_1,\ldots,X_n).$$

"Mixed discriminant" is perhaps a misnomer, since it does not vanish even if all the variables are set equal. In fact, if I is the $n\times n$ identity matrix, then $D(I,\ldots,I) = n!$.

The discriminant of $n\times n$ matrices $X_1,\ldots,X_{n^2}$ is the determinant of the $n^2\times n^2$ matrix whose i-th row is X_i, written as a $1\times n^2$ row vector. It is denoted $\Delta(X_1,\ldots,X_{n^2})$. The multiplicative property of the determinant shows that the discriminant is an invariant; the point is that if $P \in GL(n,K)$, then the linear transformation of $M_n(K)$ defined by $X \mapsto PXP^{-1}$ has determinant 1. There cannot be a unique expression for the discriminant in terms of traces, since there are many trace functions of $X_1,\ldots,X_{n^2}$ which vanish on $M_n(K)^{n^2}$. However, the expression below is a natural choice.

Theorem 9. [F2, pp. 210-211], [F6, Theorems 4 and 24]. *Let* μ *be an element of* S_{n^2} *with cycle type* $(1,3,5,\ldots,2n-1)$. *Then the discriminant* $\Delta(X_1,\ldots,X_{n^2})$ *of* $n\times n$ *matrices* $X_1,\ldots,X_{n^2}$ *is*

$$\pm \frac{1!3!5!\ldots(2n-1)!}{1!2!3!\ldots(n-1)!} \sum_{\sigma\in S_{n^2}} (\operatorname{sign}\sigma) T_{\sigma\mu\sigma^{-1}}(X_1,\ldots,X_{n^2}) .$$

There is also a well-known and useful expression for $\Delta(X_1,\ldots,X_{n^2})\Delta(Y_1,\ldots,Y_{n^2})$ in terms of traces. Let $V(X)$ be the $n^2\times n^2$ matrix whose rows are $X_1,\ldots,X_{n^2}$, written as $1\times n^2$ row vectors, and let $V^t(Y^t)$ be the $n^2\times n^2$ matrix whose columns are $Y_1^t,\ldots,Y_{n^2}^t$, written as $n^2\times 1$ column vectors. Note that the transpose Y^t of $Y \in M_n(K)$ is obtained from Y by interchanging $\binom{n}{2}$ pairs of entries, which introduces a coefficient $(-1)^{\binom{n}{2}}$ below. Thus

$$\operatorname{Det} V(X) = \Delta(X_1,\ldots,X_{n^2})$$

$$\operatorname{Det} V^t(Y^t) = (-1)^{\binom{n}{2}}\Delta(Y_1,\ldots,Y_{n^2})$$

$$V(X)V^t(Y^t) = (T(X_iY_i)),$$

and the multiplicative property of the determinant yields

$$\Delta(X_1,\ldots,X_{n^2})\Delta(Y_1,\ldots,Y_{n^2}) = (-1)^{\binom{n}{2}}\operatorname{Det}(T(X_iY_i)).$$

7. The Polynomial Identities Satisfied by $n\times n$ Matrices.

Let $K\langle X\rangle = K\langle x_1,x_2,\ldots\rangle$ be a free associative algebra over K in countably many variables, and let $K^+\langle X\rangle$ denote the ideal of polynomials with zero constant term. We egard $K^+\langle X\rangle$ as a K-algebra without unit.

A K-algebra A is said to satisfy a polynomial identity if there is a nonzero olynomial $f(x_1,\ldots,x_r) \in K^+\langle X\rangle$ such that $f(a_1,\ldots,a_r) = 0$ for all $a_1,\ldots,a_r$. The et of all $g \in K^+\langle X\rangle$ which vanish on A form an ideal called the T-ideal of identities atisfied by A. Abstractly, a T-ideal is an ideal of $K^+\langle X\rangle$ which is closed under ndomorphisms of $K^+\langle X\rangle$.

A slightly different theory arises if $K\langle X\rangle$ replaces $K^+\langle X\rangle$ in the preceding efinitions, and that is the more common choice in the literature. We have chosen $^+\langle X\rangle$, since the treatment of nil algebras in Section 11 requires it.

The ring $M_n(K)$ satisfies a polynomial identity and a great deal of effort has been evoted to the T-ideal of identities satisfied by $M_n(K)$ (see [F1]). Any T-ideal is enerated by the multilinear polynomials it contains, so it is enough to know the ultilinear polynomials satisfied by $M_n(K)$. The following theorem, which formalizes n observation of Kostant [K, Lemma 3.4] shows that the second fundamental theorem etermines the multilinear polynomial identities satisfied by $M_n(K)$.

heorem 10. *Let* $\Sigma\{a_\sigma x_{\sigma(1)}\cdots x_{\sigma(r)} \mid \sigma \in S_r\}$ *be a multilinear polynomial of degree* r *n* $K^+\langle X\rangle$. *The following are equivalent.*

1) $\Sigma\{a_\sigma x_{\sigma(1)}\cdots x_{\sigma(r)}\}$ *is a polynomial identity for* $M_n(K)$.

2) *If* $X_1,\ldots,X_{r+1} \in M_n(K)$, $\Sigma a_\sigma T(X_{\sigma(1)}\cdots X_{\sigma(r)}X_{r+1}) = 0$.

3) $\Sigma a_\sigma \sigma(12\ldots r+1)\sigma^{-1}$ *lies in* $J(n,r+1)$, *the two-sided ideal of* KS_{r+1} *generated by* *ll simple factors of* KS_{r+1} *corresponding to Young diagrams with* $\geq n+1$ *rows.*

Aside from the second fundamental theorem, only the nondegeneracy of the trace is eeded to prove Theorem 10: If $Z \in M_n(K)$, then $Z = 0$ if and only if $T(XZ) = 0$ or all $X \in M_n(K)$.

In spite of Theorem 10, many problems about the polynomial identities satisfied by $M_n(K)$ remain open. Probably the most important is whether the T-ideal of identities of $M_n(K)$ is finitely generated as a T-ideal.

. Kostant's Proof of the Amitsur-Levitzki Theorem.

Theorem 11. (Amitsur-Levitzki [A-L]). $M_n(K)$ *satisfies the standard polynomial of* *degree* 2n:

$$S_{2n}(x_1,\ldots,x_{2n}) = \sum_{\sigma\in S_{2n}} (\text{sign } \sigma)x_{\sigma(1)}\cdots x_{\sigma(2n)}.$$

Several quite different proofs of this theorem have been published. We will outline Kostant's proof [K, pp. 248-257] which was the first application of the second fundamental theorem to polynomial identities.

As a starting point, note that if $X_1,\ldots,X_{2n+1}$ are $n\times n$ matrices, then

$$T(S_{2n}(X_1,\ldots,X_{2n})X_{2n+1}) = T(S_{2n+1}(X_1,\ldots,X_{2n+1}))$$

$$= \sum_{\sigma\in S_{2n}} (\text{sign }\sigma)T(X_{\sigma(1)}\cdots X_{\sigma(2n+1)}) = \sum_{\sigma\in S_{2n}} (\text{sign }\sigma)T_{\sigma\tau\sigma^{-1}}(X_1,\ldots,X_{2n+1})$$

$$= \frac{1}{2n+1} \sum_{\sigma\in S_{2n+1}} (\text{sign }\sigma)T_{\sigma\tau\sigma^{-1}}(X_1,\ldots,X_{2n+1}),$$

where $\tau = (12\ldots 2n+1) \in S_{2n+1}$. Now Theorem 10 implies

(*) $\mathcal{S}_{2n}(x_1,\ldots,x_{2n})$ is a polynomial identity for $M_n(K)$ if and only if $\Sigma\{(\text{sign }\sigma)\sigma\tau\sigma^{-1} | \sigma\in S_{2n+1}\}$ lies in the ideal $J(n,2n+1)$ of KS_{2n+1}.

Note that $\Sigma\{(\text{sign }\sigma)\sigma\tau\sigma^{-1}\}$ is actually a central element of KA_{2n+1}, the group algebra of the alternating group. The theory of Frobenius [J-K, pp. 65-72] shows how the irreducible representations of A_r and S_r are related. The following result is a consequence of his theory.

__Theorem 12.__ _Let α be an element of S_r with cycle type $(\alpha_1,\ldots,\alpha_t)$, and let_

$$A(\alpha) = \sum_{\sigma\in S_r} (\text{sign }\sigma)\sigma\alpha\sigma^{-1}.$$

Then $A(\alpha) = 0$ unless $\alpha_1,\ldots,\alpha_t$ are distinct odd integers. If $\alpha_1,\ldots,\alpha_t$ are distinct odd integers, then $A(\alpha) \neq 0$. Furthermore, $A(\alpha) \in J(\alpha)$, where $J(\alpha)$ is the minimal two-sided ideal of KS_r corresponding to the unique self-associated Young diagram whose main hooks have lengths $\alpha_1,\ldots,\alpha_t$.

For the definitions of _self-associated_ and _main hooks_, see [J-K, p. 22, p. 67].

__Example:__ $r = 13$. There are three partitions of 13 into distinct odd parts: (13), (9,3,1), and (7,5,1). The self-associated Young diagrams having these main hook lengths are shown below.

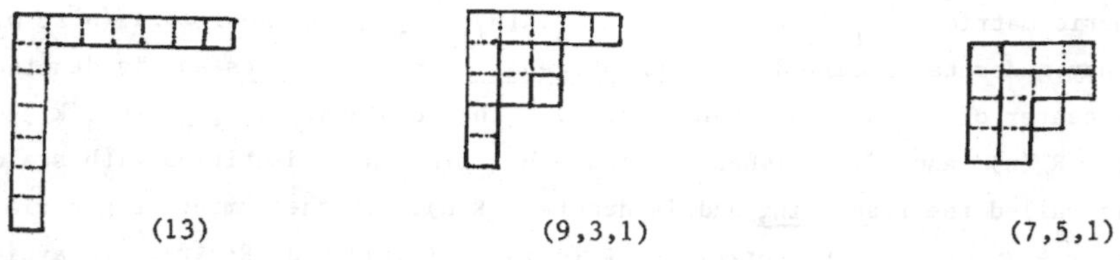

Recall that $\tau = (12\ldots 2n+1)$ and

$$A(\tau) = \sum_{\sigma \in S_{2n+1}} (\operatorname{sign} \sigma)\sigma\tau\sigma^{-1},$$

By Theorem 12, $A(\tau)$ lies in $J(\tau)$, the minimal two-sided ideal of KS_{2n+1} corresponding to the following self-associated Young diagram, a single hook of length 2n+1

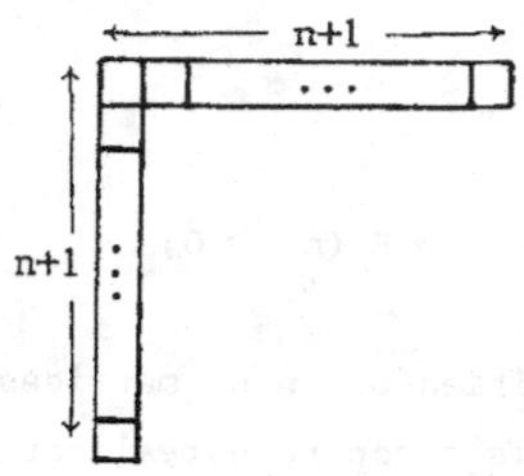

Since this Young diagram has $n+1$ rows, $J(\tau) \subseteq J(n,2n+1)$. Hence $A(\tau) \in J(n,2n+1)$, so $S_{2n}(x_1,\ldots,x_{2n})$ is a polynomial identity for $M_n(K)$, by (*). Thus the Amitsur-Levitzki Theorem is proved.

9. The Ring of Generic Matrices and the Trace Ring

Recall the generic $n \times n$ matrices $U_k = (u_{ij}(k))$ $(1 \leq i,\ j \leq n;\ \ k = 1,2,\ldots)$ whose entries are independent commuting indeterminates over K. The ring of invariants is the fixed ring

$$C(n) = K[u_{ij}(k)]^{GL(n,K)}.$$

The object of this section is to briefly describe two noncommutative rings closely related to the ring of invariants. A more complete discussion of these rings can be found in [F1, §6], [F2, §5], and [F3, §8].

The generic matrices $U_1, U_2, \ldots$ lie in $M_n(K[u_{ij}(k)])$. The K-algebra (here without unit) they generate is called the <u>ring of generic</u> $n \times n$ <u>matrices</u> and is denoted $R_0(n)$. The center of $R_0(n)$ is denoted $C_0(n)$. The subalgebra of $M_n(K[u_{ij}(k)])$ generated by $R_0(n)$ and $C(n)$ (where elements of $C(n)$ are identified with scalar matrices) is called the <u>trace ring</u> and is denoted $R(n)$. If the number of generic matrices is $r \geq 2$, we use the notations $R_0(n,r)$, $C_0(n,r)$ and $R(n,r)$. We avoid the case $r = 1$, since $R_0(n,1)$ is a polynomial ring in one variable and lacks the noncommutativity we desire.

It is easy to see that $C_0(n) = R_0(n) \cap C(n)$, which implies that there is a fiber square of ring inclusions

$$\begin{array}{ccc} R_0(n) \cap C(n) = C_0(n) & \longrightarrow & C(n) \\ \downarrow & & \downarrow \\ R_0(n) & \longrightarrow & R(n) = R_0(n)C(n). \end{array}$$

There is an exact sequence

$$0 \to M(n) \to K^+\langle X\rangle \to R_0(n) \to 0,$$

where $M(n)$ is the T-ideal of identities of $n \times n$ matrices and $K^+\langle X\rangle \to R_0(n)$ is defined by $x_k \to U_k$. Thus $R_0(n)$ is a concrete realization of the <u>relatively free ring</u> $K^+\langle X\rangle/M(n)$, which explains why $R_0(n)$ and its center arise naturally in the study of polynomial identity rings.

The trace ring $R(n)$ was originally introduced as a tool for studying the ring of generic matrices $R_0(n)$. Both rings have the same division ring of fractions, the so-called <u>generic division ring</u>, and $R_0(n)$ contains a nonzero two-sided ideal of $R(n)$. $R(n)$ is useful because it is a fairly good approximation of $R_0(n)$, but, unlike $R_0(n)$, it shares most of the desirable properties of $C(n)$. The next result shows that $R(n)$ is a close relative of $C(n)$.

<u>Theorem 13</u>. (1) [P2, § 2], [F2, Theorem 10]. <u>Let</u> $GL(n,K)$ <u>act on</u> $K[M_n(K)^r]$ <u>as usual and on</u> $M_n(K)$ <u>by</u> $X \in P^{-1}XP$, <u>where</u> $X \in M_n(K)$ <u>and</u> $P \in GL(n,K)$. (Note that P^{-1} <u>and</u> P <u>are reversed in comparison to the action on</u> $M_n(K)^r$). <u>This induces a diagonal action of</u> $GL(n,K)$ <u>on</u>

$$M_n(K) \otimes_K K[M_n(K)^r] \cong M_n(K[M_n(K)^r]),$$

<u>and</u> $R(n,r)$ <u>is the fixed algebra of this action</u>.

(2) [P2, Theorem 3.4]. $C(n,r)$ <u>is the center of</u> $R(n,r)$ <u>and</u> $R(n,r)$ <u>is a finitely generated</u> $C(n,r)$<u>-module. Hence</u> $R(n,r)$ <u>is a Noetherian ring.</u>

(3) [L], [F4, Theorem 22 and note added in proof], [T3]. <u>The multigrading of</u> $C(n,r)$ <u>by</u> $\mathbb{N}^r$ <u>extends to</u> $R(n,r)$, <u>which has a Hilbert series. For</u> $r \geq 3$, <u>the Hilbert series of</u> $R(n,r)$ <u>is a rational function which satisfies the same functional equation as the Hilbert series of</u> $C(n,r)$ <u>(Theorem 5).</u>

The map

$$f(U_1,\ldots,U_r) \to T(f(U_1,\ldots,U_r)U_{r+1})$$

is a K-vector space isomorphism between $R(n,r)$ and the K-subspace of elements of $C(n,r+1)$ having degree one in U_{r+1}-i.e. having multidegree $(\alpha_1,\ldots,\alpha_r,1) \in \mathbb{N}^{r+1}$. This implies

<u>Theorem 14. The Hilbert series of</u> $R(n,r)$ <u>and</u> $C(n,r+1)$ <u>are related by</u>

$$H[R(n,r)](t_1,\ldots,t_r) = \frac{\partial}{\partial t_{r+1}} H[C(n,r+1)](t_1,\ldots,t_{r+1})\Big|_{t_{r+1}} = 0.$$

It is very likely that $R(n,r)$ is a Cohen-Macaulay module over $C(n,r)$ and a Gorenstein ring, but neither property has been established. Finding presentations for $R(n,r)$ as a K-algebra is probably more difficult. The best result so far is a presentation for $R(3,2)$ by Le Bruyn and Van den Bergh [L-V].

10. <u>Regev's Polynomial.</u>

Let $U_1,\ldots,U_{n^2}, V_1,\ldots,V_{n^2}$ be generic $n\times n$ matrices, and define an element $F(U,V)$ in $R_0(n)$, the ring of generic $n\times n$ matrices by

$$F(U,V) = \sum_{\sigma,\rho\in S_{n^2}} (\text{sign } \sigma\rho) U_{\sigma(1)} V_{\rho(1)} U_{\sigma(2)} U_{\sigma(3)} U_{\sigma(4)} V_{\rho(1)} V_{\rho(2)} V_{\rho(3)}$$

$$\ldots U_{\sigma(n^2-2n+2)} \ldots U_{\sigma(n^2)} V_{\rho(n^2-2n+2)} \ldots V_{\rho(n^2)}.$$

In other words, U's and V's are intertwined according to the pattern $1,1,3,3,5,5,\ldots,2n-1,2n-1$. The fact that $F(U,V)$ is alternating separately in $U_1,\ldots,U_{n^2}$ and $V_1,\ldots,V_{n^2}$ implies that it is a scalar multiple of $\Delta(U)\Delta(V)I$, where $\Delta(U),\Delta(V)$ are the discriminants of $U_1,\dots,U_{n^2}$, $V_1,\dots,V_{n^2}$ respectively, and I is then $n\times n$ identity matrix. Regev [R, p.1429] conjectured that $f(U,V)$ is nonzero - i.e., the scalar is nonzero.

The main significance of the conjecture is that implies that

$$R_0(n) \supseteq \Delta(U)\Delta(V)R(n),$$

which allows quantitative information about $R_0(n)$ to be deduced from quantitative information about $R(n)$ (see [F2, pp. 212-214]).

We will outline Formanek's proof [F6] of Regev's conjecture. It provides another example of how the second fundamental theorem can be applied to polynomial identities.

By the second fundamental theorem there is a K-vector space homomorphism from KS_r into $C(n)$ whose kernel is the two-sided ideal $J(n,r)$. The image of this homomorphism is the K-vector space of multilinear invariants of $U_1,\dots,U_r$, which we denote by $C^*(n,r)$. The homomorphism $\varphi\colon KS_r \to C^*(n,r)$ induces a left and right action of S_r on $C^*(n,r)$. Neither one is the usual permutation action of S_r on $C^*(n,r)$, which is induced by the action of S_r on KS_r by conjugation. For $r \le n$, $\varphi\colon KS_r \to C^*(n,r)$ is an isomorphism, but this fact is not essential for the following construction.

Suppose that $f(U_1,\dots,U_n) \in C^*(n,n)$, and define a new invariant $\Gamma(f) \in C^*(n,n+n^2)$ (where the $n+n^2$ generic matrices are $U_1,\dots,U_n$, $V_1,\dots,V_{n^2}$) by

$$\Gamma(f)(U_1,\dots,U_n,V_1,\dots,V_{n^2})$$

$$= \sum_{\sigma\in S_{n^2}} (\operatorname{sign}\sigma) f(U_1V_{\sigma(1)},U_2V_{\sigma(2)}V_{\sigma(3)}V_{\sigma(4)},\dots,U_nV_{\sigma(n^2-2n+2)}\cdots V_{\sigma(n^2)}).$$

Since the discriminant of $V_1,\dots,V_{n^2}$ is (up to a scalar) the unique alternating multilinear function of $V_1,\dots,V_{n^2}$, standard arguments show that

$$\Gamma(f)(U_1,\dots,U_n,V_1,\dots,V_{n^2}) = \hat{f}(U_1,\dots,U_n)\Delta(V_1,\dots,V_{n^2}),$$

where $\hat{f}(U_1,\dots,U_n) \in C^*(n,n)$. Thus $f(U_1,\dots,U_n) \mapsto \hat{f}(U_1,\dots,U_n)$ defines a map $\Phi\colon C^*(n,n) \to C^*(n,n)$. It has the following property, which is surprising but easy to establish.

Theorem 15 [F6, Theorem 8]. Let $C^*(n,n)$ be a left S_n-module with the action induced by the isomorphism $\varphi: KS_n \to C^*(n,n)$. Then the map $\Phi: C^*(n,n) \to C^*(n,n)$ defined above is a homomorphism of left S_n-modules.

More generally, the above construction can be varied to give left S_r-homomorphisms from any $C^*(n,r)$ to itself.

Using Theorem 15, the map Φ can be explicitly determined and this permits an explicit calculation of the invariant

$$\Gamma(g)(U_1,\dots,U_n,V_1,\dots,V_{n^2})$$

$$= \sum_{\sigma\in S_{n^2}} (\text{sign }\sigma)T(U_1V_{\sigma(1)}U_2V_{\sigma(2)}V_{\sigma(3)}V_{\sigma(4)},\dots,U_nV_{\sigma(n^2-2n+2)}\cdots V_{\sigma(n^2)}),$$

where $g(U_1,\dots,U_n) = T(U_1\dots U_n)$. Finally, Regev's polynomial can be evaluated because its trace is a linear combination of specializations of $\Gamma(g)$. The results of Frobenius on the characters of the alternating group action play a role.

Theorem 16 [F6, Theorems 16 and 24]. Let $F(U,V) = F(U_1,\dots,U_{n^2},V_1,\dots,V_{n^2})$ be Regev's polynomial. Then

$$F(U,V) = \frac{(-1)^{n+1}}{2n-1}\left[\frac{1!3!5!\dots(2n-1)!}{1!2!3!\dots n!}\right]^2 \Delta(U)\Delta(V)I.$$

11. Nil Algebras and the Nagata-Higman Theorem

An algebra A is said to be nil of index n if $a^n = 0$ for all $a \in A$; it is nilpotent of index n if $A^n = 0$ – i.e., if $a_1\dots a_n = 0$ for all $a_1,\dots,a_n$.

A well-known theorem, now called the Nagata-Higman Theorem, asserts that nil algebras over a field of characteristic zero are nilpotent. Recently I learned from Gerald Schwarz that it has been proved ten years earlier by Dubnov and Ivanov [D-I]. Their paper was reviewed by Richard Brauer [Math. Reviews 6(1945), p. 113], but subsequently appears to have been completely overlooked, even in the Soviet Union.

Theorem 17 [D-I], [N], [H]. Let A be an algebra over a field of characteristic zero, and suppose that $a^n = 0$ for all $a \in A$. Then there is an integer $d(n)$, depending only on n, such that $A^{d(n)} = 0$.

The Nagata-Higman Theorem has a straightforward interpretation in terms of T-ideals. Let $I(x_1^n)$ denote the T-ideal of $K^+\langle X\rangle$ generated by x_1^n. Applying the theorem to the relatively free algebra $K^+\langle X\rangle/I(x_1^n)$ gives the conclusion: $x_1\dots x_{d(n)} \in I(x_1^n)$.

In [D-I] and [H] it was shown that $d(n) \le 2^n - 1$, and it follows from [Ku] that $d(n) \ge n(n+1)/2$.

The results in the sequel are applications by Razmyslov [R] and Procesi [P2], of the second fundamental theorem to nil algebras. Among them is an improved index of nilpotence in the Nagata-Higman Theorem, namely $d(n) \le n^2$. Procesi made the important observation that finding $d(n)$, the index of nilpotence of $K^+\langle X\rangle/I(x_1{}^n)$, is equivalent to finding the minimal degree of a generating set for $C(n)$, a problem which was considered in Section 5.

Several new free objects are required. Loosely speaking, the idea is to adjoint "formal traces" to $K^+\langle X\rangle$, the free algebra without unit. There is a universal construction, the <u>Hattori-Stallings Trace</u>, which assigns to any k-algebra A the K-vector space $A/[A,A]$, where $[A,A]$ denotes the K-vector subspace of A (not the ideal) generated by all $ab-ba$, $a, b \in A$. It is easy to see that

$$W = K^+\langle X\rangle/[K^+\langle X\rangle,K^+\langle X\rangle]$$

is a K-vector space with one generator for each equivalence class in $(M,\sim)$, where M is the free monoid (without 1) generated by $\{x_1,x_2,\ldots\}$ and two monomials μ and μ' are equivalent if they are cyclic permutations of one another. The <u>free pure trace ring</u> is defined to be the symmetric algebra on W. More concretely, we define

$$K[T(x)] = K[T(\mu)] = \underline{\text{free pure trace ring}}$$

to be the commutative polynomial ring (with unit) with one independent generator, denoted $T(\mu)$, for each equivalence class in $(M,\sim)$. Finally we define

$$K^+\langle T(X),X\rangle = K[T(X)] \otimes_K K^+\langle X\rangle = \underline{\text{free mixed trace ring}}.$$

Again, it is more convenient to concretely realize $K^+\langle T(X),X\rangle$ as the K-vector space spanned by all monomials

$$T(\mu_1)\ldots T(\mu_k)\mu_0,$$

where $k \ge 0$ and $\mu_0,\mu_1,\ldots,\mu_k \in M$. Note that $K^+\langle X\rangle$ is a subring of $K^+\langle T(X),X\rangle$ ($k = 0$ is allowed), but $K[T(X)]$ does not even have any elements in common with $K^+\langle T(X),X\rangle$ ($\mu_0 = 1$ is not allowed). Moreover, if Q denotes the ideal of $K^+\langle T(X),X\rangle$ generated by monomials $T(\mu_1)\ldots T(\mu_k)\mu_0$, where $k \ge 1$, then there is a split exact sequence

$$0 \to Q \to K^+\langle T(X),X\rangle \to K^+\langle X\rangle \to 0.$$

Each endomorphism of $K^+\langle X\rangle$ (a specialization of $x_1, x_2, \ldots$) induces endomorphisms f $K[T(X)]$ and $K^+\langle T(X), X\rangle$ in a natural way. We define T-ideals of $K[T(X)]$ and $^+\langle T(X), X\rangle$ to be ideals closed under such induced endomorphisms. Note that Q is a -ideal.

There is a K-linear formal trace

$$T\colon K^+\langle T(X), X\rangle \to K[T(X)]$$

efined by

$$T(T(\mu_1)..T(\mu_k)\mu_0) = T(\mu_1)...T(\mu_k)T(\mu_0).$$

e can likewise mimic the definitions of $\varphi\colon KS_r \in C(n)$ and $T_\sigma(U_1, \ldots, U_r) \in C(n)$ in ection 2 to define $\varphi\colon KS_r \to K[T(X)]$ and $T_\sigma(x_1, \ldots, x_r) \in K^+\langle X\rangle$.

The formal analogue of the fundamental trace identity is $H(x_1, \ldots, x_{n+1}) \in K[T(X)]$, here

$$H(x_1, \ldots, x_{n+1}) = \sum_{\sigma \in S_{n+1}} (\text{sign } \sigma) T_\sigma(x_1, \ldots, x_{n+1}),$$

nd the formal analogue of the multilinear Cayley-Hamilton polynomial is $^+(x_1, \ldots, x_n) \in K^+\langle T(X), X\rangle$, which is defined implicitly by

$$T(H^+(x_1, \ldots, x_n)x_{n+1}) = H(x_1, \ldots, x_n, x_{n+1}).$$

aving introduced all this formalism, we can state a theorem of Procesi which makes recise this remark in section 2: "All relations among traces are consequences of the undamental trace identity."

heorem 18 [P2, Theorem 4.5]. *Let* $\theta\colon K[T(X)] \to C(n)$ *and* $\theta\colon K^+ T(X), X \to R(n)$ *be he K-algebra homomorphisms induced by* $x_i \to U_i$ *and* $T(x_{i_1} \ldots x_{i_k}) \to T(U_{i_1} \ldots U_{i_k})$.

hen Ker θ *is generated as a T-ideal by* $H(x_1, \ldots, x_{n+1})$ *and* Ker θ^+ *is generated s a T-ideal by* $H^+(x_1, \ldots, x_n)$.

If we restrict θ^+ to $K^+\langle X\rangle \subseteq K^+\langle T(X), X\rangle$, we get the exact sequence

$$0 \to M(n) \to K^+\langle X\rangle \xrightarrow{\theta|K^+\langle X\rangle} R_0(n) \to 0$$

of section 9, where $M(n)$ is the T-ideal of identities satisfied by $M_n(K)$. Therefore $M(n) = \mathrm{Ker}\ \theta \cap K^+\langle X\rangle$, but the finite generation of $M(n)$ as a T-ideal remains open.

Let $p(x_1,\ldots,x_n) = \Sigma\{x_{\sigma(1)}\ldots x_{\sigma(n)} \mid \sigma \in S_n\}$ be the multilinearization of x_1^n. Clearly

(1) $p(x_1,\ldots,x_n)$ and x_1^n generate the same T-ideal of $K^+\langle X\rangle$, namely $I(x_1^n)$.

Recall that

$$K^+\langle T(X),X\rangle/Q \cong K^+\langle X\rangle, \tag{2}$$

where Q is the T-ideal generated by all $T(\mu_1)\ldots T(\mu_k)\mu_0 (k \geq 1)$. The formal Cayley-Hamilton polynomial has the form

$$H^+(x_1,\ldots,x_n) = p(x_1,\ldots,x_n) + q(x_1,\ldots,x_n), \tag{3}$$

where $q(x_1,\ldots,x_n)$ is a multilinear element of Q. By (1), (3) and Theorem 18, the image of $(Q + \mathrm{Ker}\ \theta^+)$ in $K^+\langle X\rangle$ is $I(x_1^n)$, so there is an isomorphism

$$K^+\langle T(X),X\rangle/(Q+\mathrm{ker}\ \theta^+) \cong K^+\langle X\rangle/I(x_1^n). \tag{4}$$

But $K^+\langle T\langle X\rangle,X\rangle/(\mathrm{Ker}\ \theta^+) \cong R(n)$, the trace ring, which is a subring of $M_n(K]u_ij(k)]$. Noting that $M_n(K[u_{ij}(k)])$ and $M_n(K)$ satisfy the same polynomial identities gives the following theorem of Procesi.

<u>Theorem 19</u> [P2, Corollary 4.8]. *If a K-algebra A satisfies the polynomial identity $x_1^n = 0$, then it satisfies all the polynomial identities satisfied by $M_n(K)$.*

Finally, we seek the minimal index $d(n)$ for the Nagata-Higman Theorem. This is the smallest integer such that $x_1\ldots x_{d(n)} \in I(x_1^n)$. The characterization of $d(n)$ uses arguments like those preceding Theorem 19, but they take place in the free trace ring in order that the map $\varphi\colon KS_{d+1} \to K[T(X)]$ can be used.

Let $K^+\langle T(X),X\rangle_d$ be the set of elements of $K^+\langle T(X),X\rangle$ which are multilinear in $x_1,\ldots,x_d$, and let

$$Q_d = Q \cap K^+\langle T(X),X\rangle_d,$$

$$(\mathrm{Ker}\ \theta^+)_d = \mathrm{Ker}\ \theta^+ \cap K^+\langle T(X),X\rangle_d$$

e the corresponding subsets of Q and $\mathrm{Ker}\ \theta^+$. Let $K[T(X)]_{d+1}$ be the set of lements of $K[T(X)]$ which are multilinear in $x_1,\ldots,x_{d+1}$. There are K-vector space somorphisms

$$K^+\langle T(X),X\rangle_d \xrightarrow{\alpha} K[T^*(X)]_{d+1} \xleftarrow{\varphi} KS_{d+1},$$

here $\alpha(f) = T(fx_{d+1})$, $\varphi(\sigma) = T_\sigma(x_1,\ldots,x_{d+1})$. Direct observation shows that

$^{-1}\alpha(Q_d) = B(d+1)$, where $B(d+1)$ is the K-subspace of KS_{d+1} spanned by all group lements except the (d+1)-cycles. The second fundamental theorem says that

$^{-1}\alpha(\mathrm{Ker}\ \theta^+)_d = J(n,d+1)$, where $J(n,d+1)$ is the sum of all simple factors of KS_{d+1} orresponding to Young diagrams with $\geq n+1$ rows. Using the isomorphism

$$K^+\langle T(X),X\rangle/(Q + \mathrm{Ker}\ \theta^+) \cong K^+\langle X\rangle/I(x_1{}^n),$$

e may therefore conclude:

$$x_1\ldots x_d \in I(x_1{}^n) \Leftrightarrow x_1\ldots x_d \in Q_d + (\mathrm{Ker}\ \theta^+)d$$

$$\Leftrightarrow \varphi^{-1}\alpha(x_1\ldots x_d) \in \varphi^{-1}\alpha(Q_d) + \varphi^{-1}\alpha(\mathrm{Ker}\ \theta^+)_d$$

$$\Leftrightarrow (1\ldots d+1) \in B(d+1) + J(n,d+1)$$

The same reasoning applies to any permutation $x_{\sigma(1)}\ldots x_{\sigma(d)}$ of $x_1\ldots x_d$. Since KS_{d+1} is spanned as a K-vector space by $B(d+1)$ and the (d+1)-cycles, we get a Theorem which is implicit in an argument of Razmyslov [R, final remark], [P3, Theorem 4.3].

<u>Theorem 20</u> <u>Let</u> $I(x_1{}^n)$ <u>be the T-ideal of</u> $K^+\langle X\rangle$ <u>generated by</u> $x_1{}^n$, <u>and let</u> $B(d+1)$ <u>and</u> $J(d+1)$ <u>be as in Theorem</u> 6. <u>Then</u> $x_1\ldots x_{d+1} \in I(x_1{}^n)$ <u>if and only if</u> $KS_{d+1} = B(d+1) + J(n,d+1)$.

The condition on KS_{d+1} is exactly the same in Theorems 6 and 20. Razmyslov also obtained the further conclusion that the index of nilpotence is the Nagata-Higman theorem can be reduced to n^2. In other words, if $A = K^+\langle X\rangle/I(x_1{}^n)$, then $A^{n^2} = 0$. Conversely, Higman [H, Theorem 2] showed that $A^d \neq 0$ if $d \leq n^2/e^2$. Kuzmin [Ku] improved this result by showing that $A^d \neq 0$ if $d < n(n+1)/2$. I thank L.A. Bokut and V.S. Drensky for bringing his article to my attention. The final paragraph of section 5 shows that the exact index of nilpotence, $d(n)$, is known only for $n \leq 3$: $d(1) = 1$, $d(2) = 3$, $d(3) = 6$.

12. Problems.

Not all of the problems below are covered by the text. Recall that

$C(n)$ = ring of invariants of $n \times n$ matrices.
$R_0(n)$ = ring of generic $n \times n$ matrices.
$R(n) = R_0(n)C(n)$ = trace ring.
$C_0(n)$ = center of $R_0(n)$.
$K^+\langle X \rangle$ = free algebra without unit.

1. What form do the first and second fundamental theorems have in characteristic $p > 0$ or over $\mathbb{Z}$?

2. What is the minimal degree for a generating set for $C(n)$? By Theorems 6 and 20, it is the same as the (best possible) index of nilpotence in the Nagata-Higman Theorem.

3. If the Hilbert series of $C(n)$ or $R(n)$ is given as a formal power series of Schur functions (power symmetric functions) is there a simple formula for the coefficient of a given Schur function (power symmetric function)?

4. Is $R(n,r)$ a Cohen-Macaulay module over $C(n,r)$? Is $R(n,r)$ a Gorenstein ring?

5. Are there canonical presentations for $C(n,r)$ and $R(n,r)$ as K-algebras, or for $R(n,r)$ as a $C(n,r)$-module?

6. If A is a subring of B, the <u>conductor</u> from B to A is the unique largest ideal of B which is contained in A. What are the conductors from $R(n)$ to $R_0(n)$ and from $C(n)$ to $C_0(n)$?

7. Is the quotient field of $C(n,r)$ a rational function field over K? (Rational function field over K = purely transcendental extension of K).

8. (Specht's problem) Is every T-ideal of $K^+\langle X \rangle$ finitely generated as a T-ideal? Is the T-ideal of identities satisfied by $M_n(K)$ finitely generated as a T-ideal?

9. What is the minimal degree of two-variable polynomial identity for $M_n(K)$?

10. An element of $K^+ X$ is a <u>central polynomial</u> for $M_n(K)$, if it takes only scalar values when evaluated on $M_n(K)$, but is not a polynomial identity for $M_n(K)$. What is the minimal degree of a central polynomial for $M_n(K)$? This is the same as the minimal degree of a nonzero element of $C(n)$.

11. Let p be a prime, $p \geq 5$. Do there exist p-th power central polynomials? Does there exist f $K^{+}\langle X\rangle$ such that f is not a central polynomial for $M_p(K)$, but f^p is a central polynomial for $M_p(K)$?

References

[A-L] S. Amitsur and J. Levitzki, Minimal identities for algebras, Proc. Amer. Math. Soc. 1(1950), 449-463.

[D] J. Dubnov, Proceedings of Seminar on Vector and Tensor Analysis, Mechanics Research Institute, Moscow State University (1935), 351-367 (Russian).

[D-I] J. Dubnov and V. Ivanov, Sur ℓ'abaissement du degre des polynomes en affineurs C.R. (Doklady) Acad. Sci. URSS 41(1943), 96-98.

[F1] E. Formanek, The polynomial identities of matrices, pp. 41-79 in "Algebraists' Homage," S.A. Amitsur et. al. Eds., Contemp. Math. Vol. 13, Amer. Math. Soc., Providence, 1982.

[F2] E. Formanek, Invariants and the ring of generic matrices, J. Alg. 89(1984), 178-223.

[F3] E. Formanek, Noncommutative invariant theory, pp. 87-119 in "Group Actions on Rings," S. Montgomery, Editor, Contemp. Math. Vol. 43, Amer. Math. Soc., Providence, 1985.

[F4] E. Formanek, Functional equations for character series associated with n×n matrices, Trans. Amer. Math. Soc. 295(1986), 647-663.

[F5] E. Formanek, Generating the ring of matrix invariants, pp. 73-82 in "Ring-Theory-Proceedings, Antwerp, 1985," F. van Oystaeyen, Editor, Lecture Notes in Math. No. 1195, Springer-Verlag, Berlin, 1986.

[F6] E. Formanek, A conjecture of Regev about the Capelli polynomial, J. Alg. (to appear).

[G] G.B. Gurevich, "Foundations of the Theory of Algebraic Invariants," P. Noordhoff Ltd., Groningen, 1964.

[H] G. Higman, On a conjecture of Nagata, Proc. Camb. Philos. Soc. 52(1956), 1-4.

[Ho] G.P. Hochschild, "Basic Theory of Algebraic Groups and Lie Algebras," Springer-Verlag, New York, 1981.

[H-R] M. Hochster and J.L. Roberts, Rings of invariants of reductive groups acting on regular rings are Cohen-Macaulay, Adv. in Math. 13(1974), 115-175.

[J-K] G. James and A. Kerber, "The Representation Theory of the Symmetric Group, Addison-Wesley, Reading, Massachusettes, 1981.

[K] B. Kostant, A theorem of Frobenius, a theorem of Amitsur-Levitski and cohomology theory, J. of Math. and Mech. 7(1958), 237-264.

[Ku] E.N. Kuzmin, On the Nagata-Higman Theorem, pp. 101-107 in "Mathematical Structure Computational Mathematics, Mathematical Modelling, Proceedings dedicated to the 60th birthday of Academician L. Iliev," Sofia, 1975 (Russian).

[L] L. Le Bruyn, The functional equation for Poincare series of trace rings of generic 2×2 matrices, Israel J. Math. 52(1985), 355-360.

[L-V] L. Le Bruyn and M. Van den Bergh, Regularity of trace rings of generic matrices, Univ. of Antwerp preprint, 1985.

[M] D. Mumford, Hilbert's fourteenth problem, pp. 431-444 in "Mathematical Developments Arising from Hilbert Problems," Proc. Symp. Pure Math. Vol. XXVIII, Part 2, F. Browder, Editor, Amer. Math. Soc., Providence, 1976.

[N] M. Nagata, On the nilpotency of nil-algebras, J. Math. Soc. Japan 4(1953), 296-301.

[P1] C. Procesi, Non-commutative affine rings, Atti Acc. Naz. Lincei, s. VIII, v. VIII, fo. 6 (1967), 239-255.

[P2] C. Procesi, The invariant theory of $n\times n$ matrices, Adv. in Math. 19(1976), 306-381.

[P3] C. Procesi, Trace identities and standard diagrams, pp. 191-218 in "Ring Theory", F. Van Oystaeyen, Editor, Lect. Notes in Math. No. 51, Dekker, New York, 1979.

[P4] C. Procesi, "A Primer of Invariant Theory," Brandeis Lecture Notes 1, Brandeis University, 1982.

[R] Y.P. Razmyslov, Trace identities of full matrix algebras over a field of characteristic zero, Izv. Akad. Nauk SSSR Ser. Mat. 38(1974), 723-756 (Russian) Translation: Math. USSR Izv. 8(1974), 727-760.

[Re] A. Regev, The polynomial identities of matrices in characteristic zero, Comm. in Alg. 8(1980), 1417-1467.

[S] K.S. Siberskii, Algebraic invariants for a set of matrices, Sib. Mat. Zhurnal 9(1)(1968), 152-164 (Russian). Translation: Sib. Math. Jour. 9(1968), 115-124.

[Sp] T.A. Springer, "Invariant Theory", Lecture Notes in Math. No. 585, Springer-Verlag, New York, 1977.

[St] R. Stanley, Hilbert functions of graded algebras, Adv. in Math. 28(1978), 57-83.

[T1] Y. Teranishi, The ring of invariants of matrices, Nagoya Math. J. 104(1986).

[T2] Y. Teranishi, Linear diophantine equations and invariant theory of matrices (to appear).

[T3] Y. Teranishi, letter of February 8, 1986.

[W] H. Weyl, "The Classical Groups," Princeton University Press, Princeton, 1946.

◆◆◆◆◆◆◆◆◆◆◆◆◆◆◆◆◆◆◆◆

FORME CANONIQUE D'UNE FORME BINAIRE

Alain Lascoux

◇◇◇◇◇◇◇◇◆◆◆◆◆◆◆◆◆◆◆◆◆◆◆◆◆◆◆◇◇◇◇◇◇◇◇

Abstract. The space of polynomials in one variable x of odd degree $2n+1$ is $2n+2$-dimensional. A linear combination of $n+1$ polynomials $(x-\beta)^{2n+1}$ takes care of the required number of parameters. Such an expression has been called the canonical form of a binary form. Sylvester has shown that this reduction is obtained through another polynomial, the catalecticant, when it has the taste of having all its roots distinct. It happens, and this is not by coincidence, that the catalecticant is an orthogonal polynomial for a certain linear functional associated to the polynomial ; the required relations are given by the theory of symmetric functions, supplemented by a little amount of divided differences and Lagrange interpolation formula. Having recovered with these tools the classical case, one can go further and treat as well the case where all the roots of the catalecticant are the same; this is what we do in Theorem 3.4 .

1. FONCTIONS SYMETRIQUES .

Etant donnés deux ensembles finis A, B d'indéterminées (dits alphabets), on note $S_j(A-B)$ les coefficients de la série rationnelle de pôles A et zéros B :

$$(1.1)\qquad \Sigma\, z^j\, S_j(A-B) = \Pi_{b\in B}\,(1-zb)\,/\,\Pi_{a\in A}\,(1-za)$$

Le polynôme dont les racines sont B s'écrit ainsi $S_n(x-B)$, avec $n = \mathrm{card}(B)$: c'est le cas où A est de cardinal 1 , $A = \{x\}$. Il est clair que

$\Pi_{b\in B}\,(1-zb)\,/\,(1-zx) = 1+\ldots+z^{n-1}\,S_{n-1}(x-B) + z^n S_n(x-B)\,/\,(1-zx)$, et donc,

$$(1.2)\qquad \text{pour tout } j\geqslant 0\ ,\quad S_{n+j}(x-B) = x^j S_n(x-B)\ .$$

La séparation du numérateur et du dénominateur de la série (1.1) obtenue en posant successivement $A=\emptyset$, puis $B=\emptyset$, donne

$$(1.3)\qquad S_j(A-B) = \Sigma_{i\in\mathbb{Z}}\ S_{j-i}(A)\,.\,S_i(-B)$$

la sommation étant en fait limitée à un nombre fini de termes, puisque pour tout $k>0$, $S_{-k}(.) = 0$. En particulier,

$$(1.4)\qquad \Pi_{b\in B}(x-b) = S_n(x-B) = x^n S_0(-B) + x^{n-1}S_1(-B) + x^{n-2}S_2(-B) + \ldots$$

et donc les $S_j(-B)$ sont les coefficients du polynôme $S_n(x-B)$ pour $0\leqslant j\leqslant n$ et sont nuls pour $j>n$.

On n'impose pas que les éléments de A et B soient distincts. Par exemple, si tous les $b \in B$ sont égaux (on écrit $B = nb$), on a $S_n(x - nb) = (x-b)^n$. Par spécialisation $b = 1$, i.e. $B = \{1, \ldots, 1\}$, on obtient

(1.5) $$S_j(-n) = (-1)^j \binom{n}{j} \quad \text{et} \quad S_j(n) = \binom{n+j-1}{j} \quad .$$

Si A est un sous-alphabet de B de cardinal m , on a

(1.6) $$S_n(x-B) \,/\, S_m(x-A) = S_{n-m}(\, x - (B-A)\,)$$

qui est bien égal à $S_{n-m}(\,(x+A) - B)$, où $x+A$ désigne l'alphabet $\{x\} \cup A$. Dans les λ-anneaux, on n'éprouve aucune réticence à additionner, soustraire ou multiplier par des réels des alphabets ; ces anneaux sont le cadre approprié aux manipulations de fonctions symétriques plus élaborées que dans ce texte.

Les présentes notations permettent d'écrire de manière condensée certaines identités binomiales. Ainsi nous aurons besoin de l'identité suivante dont on nous épargnera la vérification qui ne fait que combiner (1.3) et (1.5) .

(1.7) LEMME. Soient j, k, n des entiers $\geqslant 0$, x et y deux indéterminées. Alors $S_{k+1+j}(\, ny - (n+k)x\,) = y^{j+1-n}\,(y-x)^{k+1}\, S_{n-1}(\,(2+k+j)y - jx\,)$.

Il existe une autre manière d'écrire les polynômes, en faisant apparaître les coefficients binomiaux. Si n est un entier, E un alphabet, x une indéterminée, on a d'après (1.3) et (1.5) :

(1.8) $$S_n(E-nx) = S_n(E) - \binom{n}{1} x S_{n-1}(E) + \binom{n}{2} x^2 S_{n-2}(E) - \ldots \pm \binom{n}{n} x^n S_0(E)$$

puisque $S_j(-nx) = (-x)^j \binom{n}{j}$.

Les dérivées successives de $S_n(E - nx)$ s'écrivent commodément :

(1.9) $$\frac{d}{dx^p} S_n(\,E - nx\,) = n(n-1)\ldots(n-p+1)\; S_{n-p}(\,E - (n-p)x\,) \quad .$$

Les fonctions symétriques fondamentales sont les fonctions de Schur [Macdonald] .

(1.10) DEFINITION. Soient p un entier, $H \in \mathbb{Z}^p$, $A_1, B_1, \ldots, A_p, B_p$ des alphabets. Alors la fonction de Schur $S_H(A_1-B_1, \ldots, A_p-B_p)$ est le déterminant

$$\left| S_{h_j+j-i}(A_j - B_j) \right|_{1 \leqslant i,j \leqslant p} \quad .$$

On réserve parfois le nom de Schur au cas spécial $A_1 = \ldots = A_p = A$, $B_1 = \ldots = B_p = \emptyset$, pour lequel on simplifie la notation en se contentant de $S_H(A)$.

La théorie algébrique des polynômes orthogonaux ou des déterminants de Hankel peut s'interpréter comme l'étude du cas des partitions H rectangles , i.e. $H = (q, \ldots, q) \in \mathbb{N}^p$. Soient en effet E un alphabet (de cardinal infini!) , π la fonctionnelle (notée à droite) : $\forall\, n \geqslant 0,\ x^n \pi = S_n(E)$. Alors [Brezinski] :

(1.11) PROPOSITION. $P_n(x) = S_{n^n}(E-x) = S_{n^n 0}(E,\dots,E,x)$ <u>est le n-ième polynôme orthogonal pour la fonctionnelle</u> π .

<u>Remarque</u>. On passe de $S_{n\dots n0}(E,\dots,E,x)$ à $S_{n\dots n}(E-x)$ en multipliant le premier déterminant, à gauche, par $|S_{j-i}(-x)|$. Ainsi,

$$\begin{vmatrix} 1 & -x & 0 & 0 \\ 0 & 1 & -x & 0 \\ 0 & 0 & 1 & -x \\ 0 & 0 & 0 & 1 \end{vmatrix} \bullet \begin{vmatrix} S_3(E) & S_4(E) & S_5(E) & x^3 \\ S_2(E) & S_3(E) & S_4(E) & x^2 \\ S_1(E) & S_2(E) & S_3(E) & x \\ S_0(E) & S_1(E) & S_2(E) & 1 \end{vmatrix} = \begin{vmatrix} S_3(E-x) & S_4(E-x) & S_5(E-x) & 0 \\ S_2(E-x) & S_3(E-x) & S_4(E-x) & 0 \\ S_1(E-x) & S_2(E-x) & S_3(E-x) & 0 \\ S_0(E) & S_1(E) & S_2(E) & 1 \end{vmatrix}$$

$$= \begin{vmatrix} S_3(E)-xS_2(E) & S_4(E)-xS_3(E) & S_5(E)-xS_4(E) \\ S_2(E)-xS_1(E) & S_3(E)-xS_2(E) & S_4(E)-xS_3(E) \\ S_1(E)-xS_0(E) & S_2(E)-xS_1(E) & S_3(E)-xS_2(E) \end{vmatrix}$$

. La première forme se prête plus facilement aux calculs faisant intervenir la fonctionnelle. Ainsi $S_{3330}(E,E,E,x)\,\pi = S_{3330}(E)$ ($=0$ puisque première et dernière colonne sont identiques). La deuxième forme est due à [Cayley].

Développant $S_{n^n}(E,\dots,E,x)$ suivant sa dernière colonne, on obtient $S_{n^n}(E-x) = \Sigma_0^n (-x)^j S_{(n-1)^j n^{n-j}}(E)$, et donc, si B est l'ensemble des racines du polynôme $S_{n^n}(E-x)$, on a les relations suivantes, pour $0 \leqslant j \leqslant n$:

(1.12) $$S_{(n-1)^j n^{n-j}}(E) \,/\, S_{(n-1)^n}(E) = (-1)^{n-j} S_{n-j}(-B) .$$

Développant $S_{n^n 0}(E)$, ..., $S_{n^n\, n-1}(E)$ suivant leurs dernières colonnes, et utilisant la nullité de tous ces déterminants (qui ont chacun deux colonnes répétées) on transforme les relations (1.12) en les suivantes :

(1.13) $$S_n(E-B) = 0 = \dots = S_{2n-1}(E-B) .$$

2. DIFFERENCES DIVISEES .

Si $f(x,y,z,\dots)$ est un polynôme, la différence $f(x,y,z,\dots) - f(y,x,z,\dots)$ est divisible par $x-y$. L'opérateur, noté à droite,

(2.1) $$f(x,y,z,\dots) \longrightarrow \frac{f(x,y,z,\dots) - f(y,x,z,\dots)}{x-y} = f\,\partial_{xy}$$

est dit <u>différence divisée</u>. On peut composer de tels opérateurs ; les propriétés de l'algèbre des différences divisées sont très liées à celles du groupe symétrique (cf. [L&S 1] . Soit $B = \{\beta_0, \beta_1, \dots, \beta_n\}$ un alphabet <u>totalement ordonné</u> . On note ∂_B le produit $\partial_{\beta_0\beta_1}\,\partial_{\beta_1\beta_2} \cdots \partial_{\beta_{n-1}\beta_n}$. Si f est une fonction d'une variable, $f\,\partial_B$ désigne l'image de $f(\beta_0)$ par l'opérateur ∂_B, en considérant $f(\beta_0)$ comme une fonction en B de degré nul en $\beta_1, \ldots, \beta_n$.

(2.2) LEMME. <u>Soient B, E, F trois alphabets indépendants</u>, B <u>étant de card.</u> n, j <u>un entier</u> : $0 \leq j \leq n$. <u>Alors</u>, <u>pour tout</u> $k \geq 0$,

$$\beta_0^{\,j}\, S_k(\beta_0 + E - F)\,\partial_B = S_{k+j-n}(B+E-F) .$$

Démonstration. Tout d'abord, $\beta_0^{\,j}\, S_k(\beta_0+E-F) = S_{k+j}(\beta_0+E-F) - \beta_0^{\,j-1}\, S_{k+1}(E-F) - \ldots - \beta_0^{\,0}\, S_{k+j}(E-F)$. La série $\Sigma\, z^h\, S_h(\beta_0+E-F) = \Sigma\, z^h\, S_h(E-F)\,/\,(1-z\beta_0)$ a pour image par $\partial_{\beta_0\beta_1}$ $\quad z\,\Sigma\, z^h\, S_h(E-F)\,/\,(1-z\beta_0)\,(1-z\,\beta_1) = z\Sigma\, z^h\, S_h(\beta_0+E-F)\,/\,(1-z\beta_1)$ et donc par itération, $\Sigma z^h\, S_h(\beta_0+E-F)\,\partial_B = z^n\,\Sigma z^h\, S_h(B+E-F)$. On conclut en remarquant que l'image par ∂_B de $\beta_0^{\,j-1}\, S_{k+1}(E-F) + \ldots + \beta_0^{\,0}\, S_{k+j}(E-F)$, étant un polynôme en B de degré $j-1-n$, est nulle puisque $j \leq n$ □

L'opérateur ∂_B fournit le <u>reste</u> dans les formules d'interpolation classiques de fonctions d'une variable. On considère usuellement que $\beta_0, \ldots, \beta_{n-1}$ sont les points d'interpolation, et que $\beta_n = x$ est le point où l'on veut "approximer".

(2.3) PROPOSITION (Formule de Lagrange, cf [L&S2]). <u>Soient</u> B <u>un alphabet de cardinal</u> n+1, f <u>une fonction d'une variable à coefficients symétriques en</u> B. <u>Alors</u> $\quad \Sigma_{\beta\in B}\; f(\beta)\,/\,S_n(2\beta - B) \;=\; f\,\partial_B$.

L'expression $S_n(2\beta - B)$ n'est qu'une manière condensée d'écrire $\Pi_{\beta'\in B-\{\beta\}}(\beta-\beta')$.

3. FORME CANONIQUE

(3.1) THEOREME [Sylvester]. <u>Soient</u> $S_{2n+1}(E-(2n+1)x)$ <u>un polynôme de degré impair</u>, $B=\{\beta\}$ <u>l'ensemble des racines supposées toutes distinctes du catalecticant</u> $S_{(n+1)^{n+1}}(E-x)$. <u>Alors</u> $S_{2n+1}(E-(2n+1)x)$ <u>s'écrit sous la forme canonique</u>

$$\Sigma_{\beta\in B}\,(\beta - x)^{2n+1}\; S_n(\beta+E-B)\,/\,S_n(2\beta - B) .$$

Démonstration: la fonction $f(\beta) = (\beta-x)^{2n+1}\; S_n(\beta+E-B)$ se développe en $f(\beta) = \Sigma_{j\geq 0}\;\beta^j\, S_{n-j}(E-B)\,.\,S_{2n+1}(\beta-(2n+1)x)$ puisque $(\beta-x)^{2n+1} = S_{2n+1}(\beta-(2n+1)x)$ et que $S_n(\beta+(E-B)) = \Sigma\,\beta^j\, S_{n-j}(E-B)$. Appliquant la formule de Lagrange à $f(\beta)$, on a d'après le lemme 2.2 $\Sigma_{\beta\in B}\, f(\beta)\,/\,S_n(2\beta-B) = \Sigma_{j\geq 0}\, S_{n-j}(E-B)\,.\,S_{j+n+1}(B-(2n+1)x)$ $= S_{2n+1}(\,(E-B)+(B-(2n+1)x)\,) - S_{n+1}(E-B)\,.\,S_n(B-(2n+1)x) - \ldots - S_{2n+1}(E-B)\,.\,S_0(B-(2n+1)x)$.

Or, d'après (1.13) toutes les $S_{n+1}(E-B), \ldots, S_{2n+1}(E-B)$ sont nulles, et donc

$f(\beta_0)\,\partial_B = S_{2n+1}(\,(E-B)+(B-(2n+1)x)\,) = S_{2n+1}(E-(2n+1)x)$ □

L'expression exacte des coefficients $S_n(\beta+E-B)\,/\,S_n(2\beta-B)$ semble due à [Faa de Bruno] p.109 .

4. DEGENERESCENCE .

Le théorème précédent n'a de sens que si toutes les racines β_i du catalecticant sont distinctes, car si $\beta_0 = \beta_1$, alors $S_n(2\beta_0 - B) = (\beta_0 - \beta_1)\Pi_{i>1}(\beta_0 - \beta_i)$ $= 0 = S_n(2\beta_1 - B)$, c'est-à-dire il y a au moins deux termes infinis dans la somme $\Sigma f(\beta) / s_n(2\beta - B)$. Nous allons traiter le cas d'indistinction maximale où toutes les racines sont égales : $\beta_0 = \ldots = \beta_n = \beta$. Notre outil est la méthode dite formelle des algébristes du siècle passé, qui consiste à considérer les coefficients d'un polynôme comme les puissances successives d'une indéterminée.

En d'autres termes, étant donnés un alphabet E et une indéterminée e , on considère la fonctionnelle linéaire π , elle aussi notée à droite :

(4.1) $$\forall\ j \geqslant 0 \ , \ e^j \longrightarrow S_j(E) = e^j\,\pi \ .$$

Alors $(e-x)^{2n+1}\,\pi = S_{2n+1}(e-(2n+1)x)\,\pi = \Sigma\, S_{2n+1-j}(-(2n+1)x)\, e^j\,\pi$ = $\Sigma\, S_{2n+1}(-(2n+1)x) \,.\, S_j(E) = S_{2n+1}(E-(2n+1)x)$ qui est bien le polynôme que l'on veut étudier.

(4.2) PROPOSITION. <u>Soient</u> $S_{2n+1}(E-(2n+1)x)$ <u>un polynôme,</u> $S_{(n+1)^{n+1}}(E-x)$ <u>son catalecticant</u>. <u>Alors</u>, <u>pour que</u> β <u>soit une racine de multiplicité</u> n+1 de $S_{(n+1)^{n+1}}(E-x)$, <u>il faut et il suffit que</u> β <u>soit uneracine de multiplicité au moins</u> n+1 <u>de</u> $S_{2n+1}(E-(2n+1)x)$.

Démonstration: le polynôme $S_{2n+1}(E-(2n+1)x)$ peut s'écrire, à l'aide de (1.12), comme le déterminant $(-1)^n\, S_{1^n\, n+1}(B-E,\ldots,B-E,\ B-(2n+1)x)$. Si $B = (n+1)\beta$, alors pour tout $j \geqslant 0$, $S_{n+1+j}(B-(2n+1)x)$ est divisible par $(\beta - x)^{n+1}$; la dernière colonne du déterminant ci-dessus, a fortiori $S_{2n+1}(E-(2n+1)x)$, sont donc divisibles par $(\beta - x)^{n+1}$.

Réciproquement, s'il existe β qui soit racine de multiplicité au moins n+1 de $S_{2n+1}(E-(2n+1)x)$, alors β est racine de ce polynôme et de ses n premières dérivées, i.e. $S_{2n+1}(E-(2n+1)\beta) = 0 = S_{2n}(E-2n\beta) = \ldots = S_{n+1}(E-(n+1)\beta)$, système que l'on peut remplacer, puisque pour tout entier j , $S_j(E-\beta) = S_j(E) - \beta S_{j-1}(E)$, par le système

(4.3) $$S_{2n+1}(E-(n+1)\beta) = 0 = \ldots = S_{n+1}(E-(n+1)\beta) \ .$$

Par ailleurs, partant du déterminant $S_{(n+1)^{n+1}}(E-x) = \begin{vmatrix} S_{n+1}(E-x) & \ldots & S_{2n+1}(E-x) \\ & & \\ S_1(E-x) & \ldots & S_{n+1}(E-x) \end{vmatrix}$

et soustrayant, à l'intérieur des $S_j(E-x)$, $0, \beta, 2\beta, 3\beta, \ldots, n\beta$ dans les colonnes successives, de gauche à droite, puis dans les lignes successives de bas en haut, ce qui correspond à des combinaisons linéaires de lignes et de colonnes qui ne changent pas la valeur du déterminant, on obtient

$$S_{(n+1)^{n+1}}(E-x) = \begin{vmatrix} S_{n+1}(E-n\beta-x) & S_{n+2}(E-(n+1)\beta-x) & \dots & S_{2n+1}(E-2n\beta-x) \\ \vdots & \vdots & & \vdots \\ S_2(E-\beta-x) & S_3(E-2\beta-x) & \dots & S_{n+2}(E-(n+1)\beta-x) \\ S_1(E-x) & S_2(E-\beta-x) & \dots & S_{n+1}(E-n\beta-x) \end{vmatrix}$$

Ce dernier déterminant est triangulaire, puisque chaque terme de la partie supérieure est combinaison linéaire des polynômes (4.3). Enfin, pour ce qui est de la diagonale, $S_{n+1}(E-n\beta-x) = (e-\beta)^n(e-x)\pi = (e-\beta)^n(\beta-x)\pi$ puisque $0 = S_{n+1}(E-(n+1)\beta)$ $= (e-\beta)^n(e-\beta)\pi$; le polynôme $S_{(n+1)^{n+1}}(E-x)$ est donc divisible par $(\beta-x)^{n+1}$ □

Supposons dès lors être dans le cas totalement dégénéré, i.e. qu'il existe β, racine de multiplicité $n+1$ despolynômes $S_{2n+1}(E-(2n+1)x)$ et $S_{(n+1)^{n+1}}(E-x)$. Ce dernier polynôme est donc à un facteur près $(x-\beta)^{n+1}$ et l'on peut écrire β comme quotient de deux quelconques de ses coefficients successifs; malheureusement, ces dits coefficients sont des déterminants d'ordre $n+1$ et échappent très vite aux moyens de calculs actuels, fussent-ils aussi élaborés que le système MACSYMA.

La racine β et le quotient $Q(x) = S_{2n+1}(E-(2n+1)x)\,/\,(\beta-x)^{n+1}$ sont donnés par le théorème suivant.

(4.4) THEOREME. <u>Soit un polynôme</u> $S_{2n+1}(E-(2n+1)x)$ <u>ayant une racine</u> β <u>de multiplicité au moins</u> $n+1$, e <u>une indéterminée et</u> π <u>la fonctionnelle</u> $\{1, e, e^2, \dots\} \longrightarrow \{S_0(E), S_1(E), S_2(E), \dots\}$. <u>Alors</u>

1) $\forall\, 1 \leq j \leq n+1, \quad (e^j-\beta^j)(e-\beta)^n\pi = 0$

2) $(e-\beta)^n\pi = Q(\beta)\,/\,\binom{2n+1}{n}$

3) $((e-\beta)^n\pi)^2\binom{2n}{n} = (-1)^n S_{2n}(E-2ne)\pi$

4) $\beta = S_{2n+1}(E-2ne)\pi\,/\,S_{2n}(E-2ne)\pi = e\,S_{2n+1}(E-2ne)\pi\,/\,S_{2n+1}(E-2ne)\pi$

Démonstration:

1) est une réécriture du système (4.3) puisque pour tout $j \geq 0$, $e^j(e-\beta)^{n+1}\pi =$ $S_{j+n+1}(e-(n+1)\beta)\pi = S_{j+n+1}(E-(n+1)\beta)$ □

2) $(n+1)!\,Q(\beta)$ est égal à la valeur en $x=\beta$ de la n+1-ième dérivée de $S_{2n+1}(E-(2n+1)x)$ d'après le marquis de l'Hôpital, c'est-à-dire égal à $(2n+1)\dots(n+1)\,S_n(E-n\beta) = (2n+1)\dots(n+1)\,(e-\beta)^n\pi$ □

3) $S_{2n}(E-2ne) = S_{2n}((E-(n+1)\beta) + ((n+1)\beta-2ne)) =$

$= \Sigma_{h=0}^{n-1}\ S_{2n-h}(E-(n+1)\beta)\ S_h((n+1)\beta - 2ne) + \Sigma_{j=0}^{n}\ S_{n-j}(E-(n+1)\beta)\ S_{n+j}((n+1)\beta - 2ne).$

La première somme a tous ses termes nuls d'après (4.3). Par ailleurs, pour tout $j\geq 0$, $S_{n+j}((n+1)\beta - 2ne) = \beta^j(1-e/\beta)^n\ S_n((n+j+1)\beta - je)$ d'après (1.7). Pour calculer l'image de $S_{2n}(E-2ne)$ et $e\,S_{2n}(E-2ne)$ par π , la partie 1) du théorème nous autorise à remplacer $e^i(\beta-e)^n$, $0\leq i\leq n$, par $\beta^i(\beta-e)^n$, i.e. remplacer $S_{n+j}((n+1)\beta - 2ne)$ par $\beta^j(1-e/\beta)^n\ S_n((n+j+1)\beta - j\beta) = \beta^j(\beta-e)^n\ S_n(n+1)$.

On a donc $S_{2n}(E-2ne)\pi = \Sigma_{j=0}^{n}\ S_{n-j}(E-(n+1)\beta)\ S_n(n+1)\ \beta^j(\beta-e)^n\pi =$
$= (\beta-e)^n\pi\,.\ S_n(n+1)\ \Sigma_{j=0}^{n}\ S_{n-j}(E-(n+1)\beta)\ \beta^j = (\beta-e)^n\pi\,.\,S_n(n+1)\ S_n(E-(n+1)\beta+\beta) =$
$= (-1)^n\ S_n(e-n\beta)\pi\,.\ S_n(n+1)\ S_n(E-n\beta) = (-1)^n\ S_n(n+1)\ (\ S_n(E-n\beta)\ )^2$; on conclut à l'égalité 3) en écrivant $S_n(n+1) = \binom{2n}{n}$ □

4) Le même calcul donne $e\,S_{2n}(E-2ne)\pi = \beta\ S_{2n}(E-2ne)\pi$ et $e\,S_{2n+1}(E-2ne)\pi = \beta\ S_{2n+1}(E-2ne)\pi$; par contre, pour $k>1$, $e^k\ S_{2n}(E-2ne)\pi \neq \beta^k\ S_{2n}(E-2ne)\pi$ puisque l'on a utilisé en cours de calcul $e^j(e-\beta)^{n+1}\pi = \beta^j(e-\beta)^{n+1}\pi$, égalité qui requiert $j\leq n$ d'après le point 1) du théorème.
Finalement, on a $e\,S_{2n}(E-2ne)\pi = S_1(E).S_{2n}(E) - \binom{2n}{1}\ S_2(E).S_{2n-1}(E) + \ldots +$
$+ \binom{2n}{2n}\ S_{2n+1}(E).S_0(E) = S_{2n+1}(E-2ne)\pi$ et donc
$\beta = (\ S_{2n+1}(E-2ne)\pi\ )\ /\ (\ S_{2n}(E-2ne)\pi\ ) = (\ e\,S_{2n+1}(E-2ne)\pi\ )\ /\ (S_{2n+1}(E-2ne)\pi\ )$ □

Les deux expressions de β comme quotient de deux formes quadratiques en les coeffcients du polynôme $S_{2n+1}(E-(2n+1)x)$ sont dues à [Sondat] .

(4.5) EXEMPLE: <u>Forme cubique.</u> On suppose que le polynôme $S_3(E-3x)$ admet une racine double β et une racine simple γ : $S_3(E-3x) = (\beta-x)^2(\gamma-x)$;
on a donc $S_3(E) = \beta^2\gamma$, $S_2(E) = \beta(\beta+2\gamma)/3$, $S_1(E) = (2\beta+\gamma)/3$, ce qui implique $S_{11}(E) = (\beta-\gamma)^2/9$, $S_{12}(E) = 2\beta(\beta-\gamma)^2/9$, $S_{22}(E) = \beta^2(\beta-\gamma)^2/9$, et donc
$\beta = S_{12}(E)\,/\,2S_{11}(E) = 2S_{22}(E)\,/\,S_{12}(E)$, ce qui est bien l'expression donnée par le théorème, puisque $S_{11}(E) = S_1(E).S_1(E) - S_2(E)$, $S_{12}(E) = S_1(E).S_2(E) - S_3(E)$, $S_{22}(E) = S_2(E).S_2(E) - S_1(E).S_3(E)$.

On trouve chez [Eisenstein] une analyse de la forme cubique, avec des applications à la théorie des nombres. Le catalecticant étant $S_{22}(E) - xS_{12}(E) + x^2S_{11}(E)$, le discriminant est donc $S_{12}(E).S_{12}(E) - 4S_{11}(E).S_{22}(E)$; c'est bien l'expression d'Eisenstein aux notations près.

(4.6) EXEMPLE: Forme quintique . On suppose que l'on a une racine triple :

$S_5(E-5x) = (\beta-x)^3(\gamma-x)(\delta-x)$ et donc $S_5(E) = \beta^3\gamma\delta$, $S_4(E) = \beta^2(\beta\gamma+\beta\delta+3\gamma\delta)/5$,

$S_3(E) = (\beta^3+3\beta^2\gamma + 3\beta^2\delta + 3\beta\gamma\delta)/10$, $S_2(E) = (3\beta^2+3\beta\gamma+3\beta\delta+\gamma\delta)/10$,

$S_1(E) = (3\beta+\gamma+\delta)/5$.

On trouve, écrivant S^{ij} pour le produit $S_i(E).S_j(E)$, que

$6\,S^{22} - 8\,S^{13} + 2\,S^{04} = 3(\beta-\gamma)(\beta-\delta)/100$, $2S^{23} - 3S^{14} + S^{05} = 3\beta(\beta-\gamma)(\beta-\delta)/100$,

$6S^{33} - 8S^{24} + 2S^{15} = 3\beta^2(\beta-\gamma)(\beta-\delta)/100$ et donc

$$\beta = (2S^{23} - 3S^{14} + S^{05})/2(3S^{22} - 4S^{13} + S^{04})$$

$$= 2(3S^{33} - 4S^{24} + S^{15})/(2S^{23} - 3S^{14} + S^{05})$$

comme l'affirme le théorème.

◇◇◇◇◇◇◇◇◇◇◆◆◆◆◆◆◆◆◆◆◆◆◆◆◆◆◆◆◆◆◇◇◇◇◇◇◇◇◇◇

REFERENCES.

C.BREZINSKI - Padé-type Approximants, Birhaüser 1980

A.CAYLEY - Journal de Crelle 54(1858) 48-58, 292

J.EISENSTEIN - Journal de Crelle 27(1844) 75-79 & 89-104

FAA DE BRUNO - Théorie des formes binaires, Turin 1876

L&S 1 = A.LASCOUX & M.P.SCHÜTZENBERGER -in Invariant theory, SpringerL.N. 996

L&S 2 = A.LASCOUX & M.P.SCHÜTZENBERGER - in Séminaire d'algèbre M.P.Malliavin 1984, Springer L.N. 1146

A.LASCOUX & SHI HE - Comptes Rendus Ac.Sc.Paris, 300(1985) 681

I.G.MACDONALD - Symmetric Functions and Hall Polynomials, Oxford Mat.Mono. 1979

P.SONDAT - Nouv.Ann.Math. 19, 3ème série (1900) 25-28

J.J.SYLVESTER - Collected Work, Chelsea reprint : Tome I, p.203-216, 265-283, Tome II, p.18-27

L.I.T.P., U.E.R.Maths Paris 7
2place Jussieu, 75251 Paris Cedex05

Canonical forms for binary forms of even degree

Joseph P. S. Kung

0. Introduction

The purpose of this paper is to give a survey of the theory of canonical forms for binary forms of even degree. We shall assume familiarity with classical invariant theory for binary forms as set forth in [6].

1. Waring's problem for forms

The motivating problem in the theory of canonical forms is the following analogue of Waring's problem:

PROBLEM 1. Let $f(x_1, x_2, \ldots, x_m)$ be an m-ary form of degree n. Find the minimum number $s(f)$ such that $f(x_1, x_2, \ldots, x_m)$ can be written as a sum of nth powers of linear forms.

Like most problems in classical invariant theory, almost nothing is known for the cases $m > 2$.

For binary forms with complex coefficients, this problem was completely settled by Gundelfinger in 1885. The best way to describe Gundelfinger's solution is to use the theory of apolarity. Let

$$f(x,y) = \sum_{i=0}^{n} \binom{n}{i} a_i x^i y^{n-i} \quad \text{and} \quad g(x,y) = \sum_{j=0}^{m} \binom{m}{j} b_j x^j y^{m-j}$$

be binary forms of degree n and m, where $n \geq m$. Their *apolar covariant* $\{f,g\}$ is the binary form

$$\sum_{k=0}^{n-m} \binom{n-m}{k} c_k x^k y^{n-m-k}$$

of degree $n - m$ whose coefficients are defined by

$$c_k = \sum_{\ell=0}^{m} (-1)^{m-\ell} \binom{m}{\ell} a_{\ell+k} b_{m-\ell}.$$

Let $\mathcal{F}_k$ be the vector space of binary forms of degree k. The apolar covariant $\{f,g\}$ is a bilinear map from $\mathcal{F}_n \times \mathcal{F}_m$ to $\mathcal{F}_{n-m}$. In fact, every covariant bilinear map from $\mathcal{F}_n \times \mathcal{F}_m$ to $\mathcal{F}_{n-m}$ is a constant multiple of $\{f,g\}$. (See Lemma 5.1 in [6].)

Two binary forms $f(x,y)$ and $g(x,y)$ are said to be *apolar* if the their apolar covariant $\{f,g\}$ is identically zero. Since the apolar covariant is bilinear, the set of binary forms of degree p apolar to a given form $h(x,y)$ is a subspace of the vector space $\mathcal{F}_p$ of binary forms of degree p. We shall denote this subspace by $h^{\perp}$.

PROPOSITION 1. Let g be a nonzero form of degree m and let n be a positive integer such that $n \geq m$. Then the subspace $g^{\perp}$ in $\mathcal{F}_n$ has dimension m. Moreover, suppose

$$g(x,y) = a(\mu_1 x - \nu_1 y)^{m_1}(\mu_2 x - \nu_2 y)^{m_2} \ldots (\mu_k x - \nu_k y)^{m_k},$$

where $\mu_i x - \nu_i y$ are distinct linear factors. Then, the binary forms

$$x^j(\mu_i x - \nu_i y)^{n-j}, \quad i = 1,2,\ldots,k, \quad j = 0,1,\ldots,m_i - 1$$

form a basis for the subspace $g^{\perp}$.

PROPOSITION 2. Let

$$f(x,y) = \sum_{i=0}^{n} \binom{n}{i} a_i x^i y^{n-i}$$

and let m be a positive integer such that $m \leq n$. Then the subspace $f^{\perp}$ in $\mathcal{F}_m$ has dimension $m - r + 1$, where r is the rank of the $n - m + 1 \times m + 1$ matrix

$$[a_{i+j}]_{0 \leq i \leq n-m,\ 0 \leq j \leq m}.$$

PROPOSITION 3. Let $f(x,y)$ be a binary form of degree n. Then

(A) $f(x,y)$ can be written as the sum of m or fewer n^{th} powers of linear forms

if and only if

(B) there exists a binary form $g(x,y)$ of degree m with m distinct linear factors apolar to $f(x,y)$.

From Proposition 3, we conclude that, except on an algebraic set in $\mathcal{F}_n$, (A) is equivalent to (B); in other words, (A) and (B) are *generically* equivalent. Since (B) is technically easier to work with, we shall always consider (B) rather than (A) in this paper.

Let $f(x,y)$ be a binary form of degree n. The kth *Gundelfinger covariant* $G_k[f]$ is the $k + 1 \times k + 1$ determinant defined by:

$$G_k[f] = \det\, [\partial^{2k} f / \partial^{2k-i-j}x\, \partial^{i+j}y]_{0 \leq i,j \leq k}.$$

For example, $G_0[f]$ is the form $f(x,y)$, $G_1[f]$ is the Hessian

$$H[f] = (\partial^2 f/\partial x^2)(\partial^2 f/\partial y^2) - (\partial^2 f/\partial x\partial y)^2,$$

and

$$G_2[f] = \begin{vmatrix} \partial^4 f/\partial x^4 & \partial^4 f/\partial^3 x\partial y & \partial^4 f/\partial^2 x\partial^2 y \\ \partial^4 f/\partial^3 x\partial y & \partial^4 f/\partial^2 x\partial^2 y & \partial^4 f/\partial x\partial^3 y \\ \partial^4 f/\partial^2 x\partial^2 y & \partial^4 f/\partial x\partial^3 y & \partial^4 f/\partial^4 y \end{vmatrix}.$$

Note that $G_k[f]$ is identically zero for $k > n/2$. For $k \leq n/2$, $G_k[f]$ is a binary form of degree $(n - 2k)(k + 1)$. The forms $G_k[f]$ have the umbral representation:

$$G_k[f] = [n!/(n-2s)!(s+1)!]\langle U(f) | \prod_{i\neq j} [\alpha_i, \alpha_j]^2 \prod_i [\alpha_i, u]^{n-s}\rangle,$$

where $\alpha_1, \alpha_2, \ldots, \alpha_{s+1}$ are $s + 1$ distinct Greek umbral letters and the indices in the products runs from 1 to $s + 1$. From this, we see that $G_k[f]$ are indeed covariants of the form f.

GUNDELFINGER'S THEOREM. Let $f(x,y)$ be a binary form of degree n and let s be the minimum number such that there exists a form of degree s apolar to $f(x,y)$. Then s equals the minimum number such that the sth Gundelfinger covariant $G_s[f]$ is identically zero.

All the known results concerning representation of a binary form as a sum of powers are special cases of Gundelfinger's theorem. We shall give three examples.

SYLVESTER'S THEOREM FOR BINARY FORMS OF ODD DEGREE. Let $f(x,y)$ be a binary form of odd degree $n = 2p + 1$. Then there exists a nonzero form $g(x,y)$ of degree $p + 1$ apolar to $f(x,y)$.

PROPOSITION 4. Let $f(x,y)$ be a binary form of even degree $n = 2p$. Then there exists a nonzero form of degree $p + 1$ apolar to $f(x,y)$.

PROPOSITION 5. Let $f(x,y)$ be a binary form of degree n. Then $f(x,y) = (\mu x - \nu y)^n$ if and only if its Hessian $H[f]$ is identically zero.

Proofs of these results can be found in the classical sources [1,2,3,4,7,8,9]; modern accounts can be found in [5,6].

2. Binary forms of even degree

By Gundelfinger's theorem, $s(f) \leq p + 1$ for a binary form of even degree $n = 2p$ and $s(f) = p + 1$ for a generic form. Consider a representation of a generic form $f(x,y)$ as a sum of $p + 1$ n^{th} powers:

$$(1) \qquad f(x,y) = \sum_{j=1}^{p+1} (\alpha_j x + \beta_j y)^n .$$

Since there are $n + 1$ coefficients in $f(x,y)$, the left hand side has $n + 1$ "degrees of freedom"; however, the right hand side has $2(p + 1) = n + 2$ parameters $\alpha_1, \beta_1, \ldots, \alpha_{p+1}, \beta_{p+1}$ and hence, $n + 2$ "degrees of freedom". Hence, the representation (1) cannot be unique.

More precisely, consider the subspace $f^{\perp}$ in $\mathcal{F}_{p+1}$. By Proposition 2, it has dimension at least 2. Thus, there are infinitely many expressions of $f(x,y)$ as a sum of nth powers of linear forms. For this reason, Proposition 4 is somewhat unsatisfactory. Moreover, as long as we restrict ourselves to sums of nth powers, we shall always be faced with this defect.

We can attempt to restore uniqueness by considering canonical forms of the type

$$(2) \qquad f(x,y) = \left[\sum_{j=1}^{p} (\alpha_j x - \beta_j y)^n \right] + \lambda r(x,y),$$

where $r(x,y)$ is a binary form of degree n whose coefficients are functions of $\alpha_1, \alpha_2, \ldots, \alpha_p, \beta_1, \beta_2, \ldots, \beta_p$, and λ is a scalar.

A natural choice for $r(x,y)$ (cf. completing the square for a quadratic form) is y^n. We shall call the canonical form arising from this choice the ***constant-term canonical form***. Another canonical form, first proposed by Sylvester [8], is obtained by the choice:

$$r(x,y) = \prod_{j=1}^{p} (\alpha_j x - \beta_j y)^2 .$$

We shall call this the ***Sylvester canonical form***. For $n = 4$, the Sylvester canonical form yields a method for solving the quartic by radicals, and this is most probably its historical motivation. Besides these two canonical forms, there is a third, called the ***cyclic canonical form***.

3. Catalecticants

The technical tool for studying these canonical forms is the catalecticant. Let $f(x,y) = \sum_{i=0}^{n} \binom{n}{i} a_i x^i y^{n-i}$ be a form of even degree

$n = 2p$. The *catalecticant* $c[f]$ of the form f is the determinant of the $p+1 \times p+1$ matrix

$$[a_{i+j}]_{0 \le i,j \le p}.$$

The classical result about forms of even degree uses the catalecticant.

PROPOSITION 6. Let $f(x,y)$ be a binary form of even degree $n = 2p$. Then there exists a nonzero form of degree p apolar to $f(x,y)$ if and only if the catalecticant equals zero.

PROOF. This follows from Proposition 2 or Gundelfinger's theorem. ||

Since the apolar covariant is bilinear, the problem of reducing a form of even degree to a canonical form is generically equivalent to the following problem:

Given a form $f(x,y)$ of even degree $n = 2p$, find a nonzero form $g(x,y)$ of degree p and a constant λ such that

$$\{f(x,y) - \lambda r(x,y), g(x,y)\} \equiv 0. \tag{3}$$

Note that the coefficients of $r(x,y)$ may depend on the coefficients of $g(x,y)$.

4. The constant-term canonical form

The reduction problem for the constant-term canonical form is quite easy.

PROPOSITION 7. Let $f(x,y)$ be a binary form of even degree $n = 2p$. Then there exist a constant λ and nonzero form $g(x,y)$ of degree p such that

$$\{f(x,y) - \lambda y^n, g(x,y)\} \equiv 0$$

if and only if the catalecticant is zero, or, both the catalecticant and its 0,0-cofactor

$$t[f] = \det\, [a_{i+j}]_{1 \le i,j \le p}$$

are nonzero.

PROOF. Since

$$f(x,y) - \lambda y^n = (a_0 - \lambda)y^n + \sum_{i=1}^{n} \binom{n}{i} a_i x^i y^{n-i},$$

the catalecticant of $f(x,y) - \lambda y^n$ equals $c[f] - \lambda t[f]$.

Consider the equation

$$c[f] - \lambda t[f] = 0. \tag{4}$$

If $c[f] = 0$, then (4) can be solved by setting $\lambda = 0$. If $c[f] \neq 0$, then (4) can be solved only if $t[f] \neq 0$ by setting $\lambda = c[f]/t[f]$. ||

Since the set of binary forms f of even degree n for which $t[f] \neq 0$ is the complement of an algebraic set in the vector space $\mathcal{F}_n$, the reduction problem for the constant-term canonical form is generically solvable. Despite this, the constant-term canonical form is generally regarded as unsatisfactory because it is not preserved under linear changes of variables.

5. The Sylvester canonical form.

the Sylvester canonical form is preserved under changes of variables but its reduction problem is much more difficult. In fact, except for $n = 2$, 4, and 8, it is still unsettled.

Let $f(x,y)$ be a given binary form. To reduce f to the Sylvester canonical form, we need to find a form

$$g(x,y) = \Pi_{j=1}^{n} (\alpha_j x \cdot \beta_j y)$$

apolar to $f(x,y) \cdot \lambda r(x,y)$, where $r(x,y)$ equals $[g(x,y)]^2$. That is, we have to solve the equation

$$\{f,g\} \equiv \lambda\{g^2,g\}. \tag{5}$$

This nonlinear equation seems difficult to solve in general. In fact, for $n = 2$, it is unsolvable. However, for $n = 4$ and 8, it is equivalent to an eigenvalue problem.

We shall now discuss these two cases.

PROPOSITION 8. Let $f(x,y)$ be a binary form of degree $2p$, where $2p = 4$ or 8. Then there exists a constant λ and a form $g(x,y)$ of degree p such that $f - \lambda g^2$ is apolar to g.

PROOF. The proof depends on the following technical result.

LEMMA 1. Let $g(x,y)$ be a form of degree p, where $p = 2$ or 4. Then,

$$\{g,g^2\} = \kappa I[g]g(x,y),$$

where κ is a constant and $I[g]$ is the invariant given by the umbral representation

$$I[g] = \langle U(g) | [\alpha,\beta]^p \rangle.$$

PROOF. One can prove this by a simple but tedious calculation. An easier way is to recall the following fact (see, for example, [1, p. 44]):

If $pd - 2t$ is a non-negative even integer, the dimension of the vector space $\mathcal{C}[p;d,t]$ of covariants of degree d and order t of binary forms of degree p equals

$$\pi(p;d,t) - \pi(p;d,t-1)$$

where $\pi(p;d,t)$ is the number of partitions of t into d non-negative parts chosen from the set $\{0,1,2,\ldots,p\}$.

Using this, we conclude that $\mathscr{C}[2;3,2]$ and $\mathscr{C}[4;3,4]$ are both one-dimensional. The lemma now follows from the observation that $I[g]g$ is a non-zero covariant in $\mathscr{C}[2;3,2]$ and $\mathscr{C}[4;3,4]$.

Note that for $p = 2$, the invariant $I[g]$ is the discriminant; for $p = 4$,

$$I[g] = \{g,g\} = 2a_4a_0 - 8a_1a_3 + 6a_2^{\,2}. \quad ||$$

REMARK. For binary forms g of degree other than 2 or 4, g is not a factor of $\{g,g^2\}$. For odd degrees, an easy way to see this is to observe that if g were a factor of $\{g,g^2\}$, then $\{g,g^2\}$ must equal $\kappa I[g]g$. But $I[g]$ is identically zero if the degree is odd whereas $\{g,g^2\}$ is not identically zero.

Since $I[g]$ is a scalar, solving (5) is equivalent to solving

(6) $$\{f,g\} = \epsilon g$$

for a given form $f(x,y)$ of degree 4 or 8.

Let $f(x,y) = \sum_{i=0}^{n} \binom{n}{i} a_i x^i y^{n-i}$ be a form of even degree $n = 2p$.

Consider the linear transformation $L_f\colon \mathscr{F}_p \to \mathscr{F}_p$ defined by $L_f(g) = \{f,g\}$. Relative to the basis $\{y^p,\ pxy^{p-1},\ \binom{p}{2}x^2y^{p-2},\ \ldots\ ,\ x^p\}$, the matrix of L_f equals

$$C_f = \left[(-1)^i\binom{p}{i}a_{p+i-j}\right]_{0 \le i,j \le p}.$$

For example, if $n = 4$,

$$C_f = \begin{bmatrix} a_4 & -2a_3 & a_2 \\ a_3 & -2a_2 & a_1 \\ a_2 & -2a_1 & a_0 \end{bmatrix}.$$

Because the determinant of the matrix C_f is a constant multiple of $c[f]$, we call C_f the ***catalecticant matrix*** of the form f. To solve (2) is equivalent to finding an eigenvalue and eigenvector for C_f: indeed, if ϵ is an eigenvalue for C_f and $(b_0,b_1,\ \ldots\ ,b_p)^t$ an eigenvector belonging to ϵ, then ϵ and $g(x,y) = b_px^p + pb_{p-1}x^{p-1}y + \ \ldots\ + b_0y^p$ form a solution to (6). Note that λ is related to ϵ by the equation

$$\epsilon = \kappa I[g]\lambda.$$

This completes the proof of Proposition 8. ||

Note that generically, there are $p + 1$ distinct eigenvalues and eigenvectors for C_f, and hence, $f(x,y)$ can be put into Sylvester canonical form in $p + 1$ different ways. Thus, uniqueness is not quite restored.

We conclude §5 by stating what is perhaps the main open problem in the theory of canonical forms for binary forms.

PROBLEM 2. For which even integers is the reduction problem for the Sylvester canonical form generically solvable?

6. Solving the quartic equation by radicals

Proposition 8 yields a method for solving the quartic polynomial equation by radicals. We shall describe this method by working out a numerical example.

Consider the quartic polynomial

$$p(x) = 20x^4 - 28x^3 + 18x^2 - 104x + 40.$$

Let $f(x,y) = y^4 p(x/y)$. The normalized coefficients of $f(x,y)$ are

$$a_0 = 40,\ a_1 = -26,\ a_2 = 3,\ a_3 = -7,\ a_4 = 20.$$

Hence, the catalecticant matrix C_f equals

$$\begin{bmatrix} 3 & 52 & 40 \\ -7 & -6 & -26 \\ 20 & 14 & 3 \end{bmatrix}$$

and its characteristic equation is

$$\epsilon^3 - 99\epsilon + 24030 = 0.$$

Solving this by radicals (using, say, the method in [6, p.72]), we obtain -30, $15 + 24i$, $15 - 24i$ (where $i = \sqrt{-1}$) as the eigenvalues of C_f. (Note that because the trace of C_f is zero, the three eigenvalues always sum to zero.) We shall use the real eigenvalue -30. The eigenvector associated with -30 is $(-2,1/2,1)^t$. Thus, $\lambda = 10$ and

$$g(x,y) = x^2 + xy - 2y^2 = (x + 2y)(x - y).$$

Let $X = x + 2y$ and $Y = x - y$. By Proposition 1,

$$f(x,y) - 10X^2Y^2 = aX^4 + bY^4.$$

Solving for a and b, we obtain

$$f(x,y) = (-2/3)[X^4 - 15X^2Y^2 - 16Y^4].$$

Since the left hand side is a quadratic in X^2 and Y^2, we can factor it by radicals, thus obtaining

$$\begin{aligned} f(x,y) &= (-2/3)(X^2 - 16Y^2)(X^2 + Y^2) \\ &= (-2/3)(X + 4Y)(X - 4Y)(X + iY)(X - iY) \\ &= (-2/3)(5x - 2y)(-3x + 6y)((1+i)x + (2-i)y)((1-i)x + (2+i)y). \end{aligned}$$

We conclude that the roots of $p(x)$ are $2/5$, 2, $(-1 + 3i)/2$,

$(-1 - 3i)/2$.

We remark that if the quadratic $g(x,y) = (\mu x - \nu y)^2$ is the square of a linear form, then its discriminant is zero. Hence, $\lambda = 0$ and f is apolar to g. By (2.3), $f(x,y) = (\mu x - \nu y)^3(\alpha x + \beta y)$ and can be easily factored.

7. The cyclic canonical form for the sextic

The cyclic canonical form is defined only for sextic forms. It is obtained by choosing

$$r(x,y) = [\prod_{i=1}^{3} (\alpha_i x - \beta_i y)][\prod_{i=1}^{3} \{(\alpha_i x - \beta_i y) - (\alpha_i x - \beta_i y)\}],$$

where subscripts are reduced modulo 3. Writing $U = (\alpha_i x - \beta_i y)$, $V = (\alpha_i x - \beta_i y)$, $W = (\alpha_i x - \beta_i y)$, this choice of $r(x,y)$ yields the canonical form

$$f(x,y) = U^6 + V^6 + W^6 + \lambda UVW(U - V)(V - W)(W - U).$$

The interest in the cyclic canonical form lies in the fact that its reduction problem is generically solvable: in fact, the reduction can be accomplished using the eigenvalue method in §6. To prove this, it suffices to prove the following lemma.

LEMMA 2. Let U, V, and W be linear forms. Then,

$$\{UVW, UVW(U - V)(V - W)(W - U)\} = \kappa UVW,$$

where κ is a scalar function depending the coefficients of U, V, and W.

PROOF. Let $g(x,y) = UVW$. Since g is a cubic, we can apply Sylvester's theorem to write: $g = X^3 + Y^3$, where X and Y are linear forms. Renaming and multiplying by constant factors if necessary, we have

$$U = X + Y,\ V = \omega(X + \omega Y),\ W = \omega^2(X + \omega^2 Y),$$

where ω is a primitive cube root of unity. By a routine computation,

$$(U - V)(V - W)(W - U) = (1 - \omega)^3(X^3 - Y^3).$$

Hence,

$$\begin{aligned}\{UVW, UVW(U - V)(V - W)(W - U)\} &= (1 - \omega)^3\{X^3 + Y^3, X^6 - Y^6\} \\ &= (1 - \omega)^3(X^3 + Y^3) = (1 - \omega)^3 UVW.\end{aligned}$$ ||

The proof of Lemma 3 depends heavily on the fact that a generic cubic can be written as the sum of two cubics: thus, it is unlikely that there is an extension of the cyclic canonical form to higher degrees.

Acknowledgement. The work in this paper was partially supported by a North Texas State University Faculty Research Grant.

References

1. L. E. Dickson, *Algebraic invariants*, Wiley, New York, 1914.
2. E. B. Elliott, *An introduction to the algebra of quantics*, 2nd Edition, Oxford Univ. Press, Oxford, 1913.
3. J. H. Grace and A. Young, *The algebra of invariants*, Cambridge Univ. Press, Cambridge, 1903.
4. S. Gundelfinger, Zur Theorie der binären Formen, *J. Reine Angew. Math*. 100(1886), 413-424.
5. J. P. S. Kung, Gundelfinger's theorem for binary forms, preprint.
6. J. P. S. Kung and G.-C. Rota, The invariant theory of binary forms, *Bull. Amer. Math. Soc.* (*New Series*) 10(1985), 27-85.
7. J. J. Sylvester, *An essay on canonical forms, supplement to a sketch of a memoir on elimination*, George Bell, Fleet Street, 1851 (= *Collected mathematical papers*, Vol. I, Paper 34).
8. J. J. Sylvester, On a remarkable discovery in the theory of canonical forms and of hyperdeterminants, *Math. Mag.* 2(1851), 391-410 (= *Collected mathematical papers*, Vol. I, Paper 41).
9. J. J. Sylvester, *Collected mathematical papers*, Vol. I-IV, Cambridge Univ. Press, Cambridge, 1904-1912.

Department of Mathematics
North Texas State University
Denton, Texas 76203

Invariant Theory and Differential Equations

Peter J. Olver†
School of Mathematics
University of Minnesota
Minneapolis, MN
USA 55455

1. Introduction. Recent years have witnessed a resurgence of interest in classical invariant theory. In the field of differential equations, it has become of steadily increasing importance, not only in the applications to be discussed here, but also in the theory of canonical forms of Hamiltonian systems, [4], and the study of conservation laws, [20]. The present work originally arose in the study of nonconvex variational problems of interest in elasticity for which one could prove existence of weak minimizers. A transform introduced by Gel'fand and Dikii, [7], and Shakiban, [19], changes this problem into one about the primality of certain determinantal ideals, and thus provides a complete solution. Subsequently the transform method has been applied to a wide range of problems arising in the study of differential equations and the calculus of variations. It has also been recognized, [14], that when the functions are homogeneous polynomials, the transform method is equivalent to the classical symbolic method of invariant theory. This paper will review the transform method, its relationship and application to classical invariant theory, and its application to problems arising in the calculus of variations. The last section provides a brief summary of a new, and potentially important theory of higher order differential forms ("hyperforms") which has arisen in an attempt to understand new divergence identities for transvectants which are a direct result of these investigations.

It is a pleasure to thank Frank Grosshans and the orgainzing committe for inviting me to present these results in the conference. I hope that this survey paper will motivate other researchers in invariant theory to seriously look at the transform theory and its applications.

† *Research Supported in Part by NSF Grant MCS 81-00786 .*

Throughout, we will let $x=(x^1,\ldots,x^p)$ be the independent variables and $u=(u^1,\ldots,u^q)$ be the dependent variables in some system of differential equations, so that the u's are to be viewed as functions of the x's. Partial derivatives will be denoted by subscripts, e.g. $u_{12}=\partial^2 u/\partial x^1\partial x^2$, or more generally using multi-index notation, so u^α_I for $I=(i_1,\ldots,i_k)$ will denote the k-th order partial derivative $\partial^k u^\alpha/\partial x^{i_1}\ldots\partial x^{i_k}$. A differential polynomial is a complex-valued polynomial in the derivatives u^α_I (for simplicity we are excluding explicit x dependence in our differential polynomials). A differential polynomial is called <u>differentially homogeneous</u> of order k if it depends exclusively on derivatives of order k; <u>algebraically homogeneous</u> refers to the usual concept of homogeneity for polynomials. Thus $u_{11}u_{22}-u_{12}$ is differentially homogeneous of degree 2, but not algebraically homogeneous, whereas $u_1u_{22}-u_2u_{11}$ is algebraically homogeneous of degree 2, but not differentially homogeneous. We let D_iP denote the <u>total derivative</u> of the differential polynomial P with respect to x^i, meaning that we differentiate P treating the u's as functions of the x's. For instance, $D_2(u_1u_{23}) = u_{12}u_{23} + u_1u_{223}$. The <u>total divergence</u> of a p-tuple $P=(P_1,\ldots,P_p)$ of differential polynomials is the differential polynomial

$$\mathrm{Div}\,P = D_1P_1+\ldots+D_pP_p.$$

The following questions are of importance for applications:

1) Characterize all differentially homogeneous differential polynomials Q which can be written as divergences: $Q = \mathrm{Div}\,P$ for some p-tuple P. An important example is the Jacobian determinant

$$\partial(u,v)/\partial(x^1,x^2) = u_1v_2 - u_2v_1 = D_1(uv_2) + D_2(-uv_1).$$

2) Characterize all differentially homogeneous <u>null divergences</u>, meaning p-tuples of differential polynomials satisfying the identity $\mathrm{Div}\,P=0$ for <u>all</u> functions u(x). An example is the "Jacobian identity"

$$D_1(u_2v_3 - u_3v_2) + D_2(u_3v_1 - u_1v_3) + D_3(u_1v_2 - u_2v_1) = 0.$$

3) More generally, characterize all <u>non-negative divergences</u>, meaning p-tuples of differential polynomials satisfying the identity $\mathrm{Div}\,P\geq 0$ for all u.

4) Characterize all differentially homogeneous differential polynomials Q which can be written as <u>higher order divergences</u>: $Q = \mathrm{Div}^k P$. More explicitly, we want to write

$$Q = \sum D_I P_I$$

for certain differential polynomials P_I. Here the sum is over all k-th order multi-indices $I=(i_1,\ldots,i_k)$, $1 \le i_\nu \le p$, with D_I denoting the corresponding k-th order total derivative $D_{i_1}\ldots D_{i_k}$. An important example is the Hessian of a function u, which is a second order divergence:

$$u_{11}u_{22} - u_{12}^2 = D_1^2(-u_2^2) + D_1 D_2(u_1 u_2) + D_2^2(u_1^2).$$

Problem 1 arises in Ball's theory of polyconvex variational problems, which are of great interest in elasticity; the solution is essentially that all such homogeneous divergences are linear combinations of Jacobian deteminants, [2]. Problem 2 arises in the classification of conservation laws of partial differential equations, where the null divergences are known as trivial conservation laws or, ocassionally, strong conservation laws since they hold for all functions u; the solution is that all such p-tuples are linear combinations of certain natural generalizations of the basic Jacobian identity given above, [16]. Problem 3 arises in the theory of continuum thermomechanics, where the Coleman-Noll procedure, [3], applied to the basic inequality arising from the second law of thermodynamics, results in such divergence inequalities; here the solution is essentially that any non-negative divergence must actually be a null divergence, and so problem 3 reduces to problem 2, [17]. This result was recently used by Dunn and Serrin, [6], in their theory of interstitial working. Finally, problem 4, which is the most interesting from the point of view of classical invariant theory, arose in generalizations of the applications of problem 1 to the variational problems of elasticity, and was used to produce nonconvex variational problems with rather weak coercivity conditions for which it was still possible to prove the existence of weak minimizers, [14]. The solution to this last problem, to be explained in more detail below, is that such a differential polynomial must be a linear combination of k-th order <u>transvectants</u> of the functions u and their derivatives, the Hessian being a multiple of the second order transvectant $(u,u)^{(2)}$, in the case that u is a homogeneous polynomial function.

2. The Transform. The key to the solution of the above problems is the introduction of a transform which, like the Fourier transform of classical analysis, changes questions about derivatives and differential polynomials into questions about ordinary algebraic polynomials, thus making them amenable to the powerful techniques of commutative algebra and invariant theory. A special case of this transform was introduced by Gel'fand and Dikii, [7], in connection with the Korteweg-deVries equation and the formal calculus of variations. It was generalized by Shakiban, [19], [20], and used to apply the invariant theory of finite groups to the study of conservation laws of differential equations. The present version is essentially the same as that discussed by Ball, Currie and Olver, [2], in the solution of the first and fourth problems of section 1. Subsequently, [14], this transform was, in the special case of polynomial functions u, recognized to be equivalent to the standard symbolic method of classical invariant theory.

In order to introduce the transform, it is easiest to start with the case when there is just one dependent variable u (so q=1), depending on p independent variables. Consider an algebraically homogeneous differential polynomial P of degree r (but not necessarily differentially homogeneous). Its transform, $\hat{P} = \mathcal{F}(P)$, will be an algebraic polynomial $\hat{P}(Z)$ of the r×p matrix of independent variables $Z = (z_j^i)$, $1 \le i \le p$, $1 \le j \le r$. The explicit formula for $\hat{P}$ is determined as follows: If P is a linear differential polynomial, so r=1, then $Z = (z^1, \dots, z^p)$ is a single row vector, and $\hat{P}(Z)$ is the ordinary Fourier transform $\hat{P}(k)$ of P when $z^j = \sqrt{-1}\, k^j$. For example, if $P = u_{11} + u_{22}$, then $\mathcal{F}(P) = (z^1)^2 + (z^2)^2$. In general, $\mathcal{F}(u_I) = z^I$, where, for $I = (i_1, \dots, i_k)$, $z^I = z^{i_1} \cdot \ldots \cdot z^{i_k}$. For higher degree differential polynomials, a natural first try for $\mathcal{F}$ would be to Fourier transform each derivative of u using a different row of the matrix Z to distinguish them; in other words, try $\mathcal{F}(u_{I_1} \cdot \ldots \cdot u_{I_r}) = (z_1)^{I_1} \cdot \ldots \cdot (z_r)^{I_r}$. However, this is ambiguous, since we can commute the u_{I_ν}'s. However, this ambiguity can be easily resolved by introducing the <u>symmetrizing map</u> $\sigma = (r!)^{-1} \Sigma\, \pi$, where the sum is over all permutations π of the integers 1,...,r, and where $\pi(Z) = (z^i_{\pi(j)})$, $\pi[\hat{P}(Z)] = \hat{P}(\pi(Z))$. Thus, define the <u>transform</u> of an r-th degree differential monomial to be

$$\mathcal{F}(u_{I_1} \cdot \ldots \cdot u_{I_r}) = \sigma\{(z_1)^{I_1} \cdot \ldots \cdot (z_r)^{I_r}\}$$

and extend $\mathcal{F}$ by linearity.

Example. For the Hessian, $P = u_{11}u_{22} - u_{12}^2$, we have p=r=2, so Z is a 2×2 matrix of variables, which for simplicity we denote by

$$Z = \begin{bmatrix} z^1 & z^2 \\ w^1 & w^2 \end{bmatrix}.$$

We find that

$$\begin{aligned} \mathcal{F}(P) &= \sigma\{(z^1)^2(w^2)^2 - z^1z^2w^1w^2\} \\ &= \tfrac{1}{2}\{(z^1)^2(w^2)^2 + (z^2)^2(w^1)^2 - 2z^1z^2w^1w^2\} \\ &= \tfrac{1}{2}(\det Z)^2 . \end{aligned}$$

Theorem. The transform $\mathcal{F}$ determines a linear isomorphism from the space of algebraically homogeneous differential polynomials to the space of symmetric algebraic polynomials of the matrix of variables Z. (By definition, $\hat{P}(Z)$ is <u>symmetric</u> if $\pi(\hat{P}) = \hat{P}$ for all permutations π.)

More generally, if there is more than one dependent variable, so $q>1$, then we define the transform on differential polynomials which are homogeneous of degree r_α in u^α, where $r=r_1+\dots+r_q$, simply by writing the u's in each differential monomial in ascending order, and, instead of using the full symmetrizing map σ, just using $\tilde{\sigma} = (\prod r_\nu!)^{-1}\cdot\sum' \pi$, the sum now being only over those permutations which permute each set of r_ν rows of Z among themselves. For instance, if P depends quadratically on derivatives of u^1 and linearly on derivatives of u^2, then $r=3$, and the only two permutations occuring in $\tilde{\sigma}$ are the identity and (12). The isomorphism theorem, with the proper interpretation of "symmetric", works just as before. (This is equivalent to, but slightly different from, the procedure used in [2], [14].)

The key to the utility of the transform method is its ability to change differential operations into algebraic operations. Two particularly important operations are the total derivative and the Euler operator, or variational derivative, E from the calculus of variations, cf. [7]. If L is any linear operator on the space of differential polynomials, then we let $\hat{L}$ be the corresponding "transformed" operator on the space of symmetric polynomials $\hat{P}(Z)$.

Proposition. The transforms of the total derivative D_j and the Euler operator are given by

$$\hat{D}_i P = (z_1^i + \dots + z_r^i)\cdot\hat{P}(Z),$$

and, letting z_j denote the j-th row of Z,

$$\hat{E}(\hat{P}) = \hat{P}(z_1,\dots,z_{r-1},-z_1-\dots-z_{r-1}).$$

(Note that E(P) has algebraic degree one less than that of P.)

As an immediate corollary, we derive the well-known result, cf. [7], that the kernel of the Euler operator is the image of the total divergence:

$E(P)=0$ if and only if $P = \mathrm{Div}\, Q$ for some p-tuple Q.

Indeed, transforming the above statement, we see that it is equivalent to the trivial algebraic proposition

$\hat{P}(z_1,\dots,z_{r-1},-z_1-\dots-z_{r-1})=0$ if and only if $\hat{P}(Z) = \sum_i (z_1^i + \dots + z_r^i)\cdot\hat{Q}_i$

for polynomials $\hat{Q}_i(Z)$.

3. The Symbolic Method. In the special case when each function $u^\alpha = \sum \binom{m}{I} c_I^\alpha x^I$ is a homogeneous polynomial or form of degree m, the differential polynomial P will evaluate to a polynomial function of the coefficients c_I^α as well as the x's. As such, it will have an umbral representation determined by the classical symbolic method, cf. [8], [9], [10]. Here we will follow Gurevich's notation for symbolic factors of the first and second kind, as the convention of Kung and Rota in which they are both brackets is rather special to the case of binary forms (p=2). Thus, if $z_j=(z_j^1,\dots,z_j^p)$, $j=1,\dots,r$, are umbral letters (the reason we use z rather than the more standard Greek letters will be clear presently), we use the notation

$$(z_j x) = z_j^1 x^1 + \dots + z_j^p x^p$$

for the symbolic factors of the first kind, and

$$[z_{j_1},\dots,z_{j_p}] = \det(\, z_{j_\nu}^i) \equiv \det(Z_J)$$

for the symbolic factors of the second kind, or <u>bracket factors</u>. Note that if we form an r×p matrix Z out of the letters $z_1,\dots,z_r$, then the bracket factor $[z_{j_1},\dots,z_{j_p}]$ can be identified with the determinant of the p×p minor Z_J consisting of rows $j_1,\dots,j_p$ of Z.

A key observation is that, apart from inessential symbolic factors of the first kind and a multiplicative constant, the transform of P agrees with the unique symmetric umbral representation of P. Specifically, we have the following result.

Theorem. Let P[u] be a differential polynomial, which is both differentially homogeneous of degree k and algebraically homogeneous of degree r_α in the k-th derivatives of u^α, with $r=r_1+\dots+r_q$. Let $\hat{P}(Z)$ be the transform of P. Let $u^\alpha(x)$ be (different) homogeneous m-th order polynomial functions, and let P(x) denote the evaluation of P when $u^\alpha=u^\alpha(x)$. Then the function

$$\hat{P}(Z)\cdot\{m!/(m-k)!\}^r\cdot\prod_{j=1}^{r}(z_jx)^{m-k}. \qquad (*)$$

is the symmetric umbral representation of $P(x)$, in which we are viewing the first r_1 rows of Z as sets of equivalent letters corresponding to u^1, the second r_2 rows as equivalent letters corresponding to u^2, etc., and (z_jx) is the symbolic factor of the first kind corresponding to the letter z_j which forms the j-th row of Z.

Conversely, if $\hat{Q}(Z)$ is a symmetric umbral representative of a polynomial $Q(x)$ depending on a collection of forms $u^1,\ldots,u^q$, and the only appearance of x in $\hat{Q}$ is through $(m-k)$-th powers of symbolic factors of the first kind, then there is an equivalent transform polynomial $\hat{P}(Z)$ related to $\hat{Q}$ by (*) such that if $P[u]$ is the corresponding differential polynomial, then P has evaluation $P(x)=Q(x)$.

This readily generalizes to the case when P is not differentially homogeneous, or when the forms u^α have different degrees, in which case the symbolic factors of the first kind occur with different powers; we leave the general statement to the interested reader.

This theorem gives an effective and simple procedure for answering problem 7 in Kung and Rota's paper, [10], on how to find the differential polynomial corresponding to such a symbolic expression. This is done by first ignoring the inessential symbolic factors of the first kind, second dividing through by the appropriate product of factorials, and third undoing the transform $\mathfrak{F}$. This last step is also quite elementary to accomplish. (There are, of course, other symbolic expressions depending on x and the coefficients of the forms which cannot not be written soley in terms of symbolic factors of the first kind, but a) they are never invariants, and b) they cannot be written as differential polynomials which do not explicitly depend on x.)

Example. The Hessian of a binary form degree m, i.e. $u_{11}u_{22}-u_{12}^2$, has the umbral representation

$$\hat{H}=\tfrac{1}{2}m^2(m-1)^2[zw]^2(zx)^{m-2}(wx)^{m-2},$$

cf. [10; page 70]. Comparing with (*), we see that the corresponding transform polynomial is just

$$\hat{P}=\tfrac{1}{2}[zw]^2=\tfrac{1}{2}(\det Z)^2,$$

where Z is the above 2×2 matrix. Thus we recover the standard representation of the Hessian of a form as a polynomial in its second order derivatives.

As a second example, consider the transvectant $(u,u)^{(4)}$, which has umbral representation

$$\hat{T} = \tfrac{1}{2}[zw]^4(zx)^{m-4}(wx)^{m-4}.$$

The corresponding transform polynomial is, according to (*),

$$\hat{P} = \tfrac{1}{2}(m-4)!^2\cdot m!^{-2}\cdot[zw]^4 = \tfrac{1}{2}(m-4)!^2\cdot m!^{-2}\cdot(\det Z)^4$$
$$= \tfrac{1}{2}(m-4)!^2\cdot m!^{-2}\cdot\{(z^1)^4(w^2)^4 - 6(z^1)^3z^2w^1(w^2)^3 + 10(z^1)^2(z^2)^2(w^1)^2(w^2)^2 - 6z^1(z^2)^3(w^1)^3w^2 + (z^2)^4(w^1)^4\},$$

It is easy to see that this is the transform of the differential polynomial

$$P = (m-4)!^2\cdot m!^{-2}\cdot\{u_{1111}u_{2222} - 6u_{1112}u_{1222} + 5u_{1122}^2\}.$$

If we had done the transvectant $(u,v)^{(4)}$ of two different forms of degree m, then we would recover the classical formula

$$(m-4)!^2\cdot m!^{-2}\cdot\{u_{1111}v_{2222}-6u_{1112}v_{1222}+10u_{1122}v_{1122}-6u_{1222}v_{1112}+u_{2222}v_{1111}\},$$

cf. [8; page 227].

This clearly demonstrates the power of the transform method for determining the differential polynomial expressions of formulas from classical invariant theory.

In fact, the transform approach to classical invariant theory accomplishes more. Since it works for arbitrary smooth (C^∞) functions, not just homogeneous polynomials, it leads immediately to an <u>invariant theory of smooth (or analytic) functions</u>. In this case, the invairants to be considered are differential polynomials in the functions, which are unchanged (up to a factor) under the action of the general linear group GL(p). It is not difficult to see that the induced action of GL(p) on the space of differential polynomials commutes with the transform. Consequently, the First Fundamental Theorem of Classical Invariant Theory, [9], [10], immediately provides a classification of all invariant differential polynomials.

Theorem. A differential polynomial P[u] is an invariant under the general linear group GL(p) if and only if its transform $\hat{P}(Z)$ is a polynomial in the symbolic factors of the second kind, i.e. a <u>bracket polynomial</u>. In particular, this requires that r, the degree of algebraic homogeneity of P, be greater than or equal to p, the number of independent variables, and that $\hat{P}$ be a polynomial in the determinants of the p×p minors of the r×p matrix Z.

There are, of course, relations among the invariant differential polynomials, all stemming from the basic syzygy among the brackets:

$$\sum_{j=1}^{p+1} (-1)^j [z_1,\ldots,z_{j-1},z_{j+1},\ldots,z_{p+1}]\cdot[z_j,w_1,\ldots,w_{p-1}] = 0.$$

However, if the functions u^α are homogeneous polynomials, or, more generally, homogeneous functions of x, then there are additional relations stemming from the syzygy between the brackets and the symbolic factors of the first kind

$$\sum_{j=1}^{p+1} (-1)^j [z_1,\ldots,z_{j-1},z_{j+1},\ldots,z_{p+1}]\cdot(z_j x) = 0,$$

even though the latter factors do not explicitly appear in the transform representation of P. We do not have space to go into the full details of the interplay between the invariant differential polynomials and the classical invariants of homogeneous polynomials, but there are three points to note. First, there are far more invariant differential polynomials than classical invariants because we only have "half" the number of syzygies at our disposal in the former case. In particular, the invariant differential polynomials are not an algebraically finitely generated ideal. Second, the first fact immediately leads to a number of "universal identities" which are valid for arbitrary homogeneous functions, all of which come from differential polynomials which give the same classical invariant when evaluated on homogeneous polynomial functions. The simplest example is the invariant differential polynomial whose transform is $[z_1z_2]^2[z_1z_3][z_2z_3]$, but whose corresponding umbral representation when u is a form always vanishes. The differential polynomial has the explicit formula

$$u_{11}(u_{112}u_{222}-u_{122}^2) - u_{12}(u_{111}u_{222}-u_{112}u_{122}) + u_{22}(u_{111}u_{122}-u_{112}^2) = 0,$$

which vanishes for all homogeneous functions $u(x^1,x^2)$, but is certainly not zero for all smooth functions u. Finally, the classical invariant-theoretic concept of a perpetuant, [8], [10], which is usually referred to as "an invariant of a form of infinite degree", can be reinterpreted in this light as an invariant differential polynomial of a homogeneous function of non-integral degree of homogeneity, i.e. $u(\lambda x) = \lambda^\alpha \cdot u(x)$, where α is not an integer (in particular, u is not a polynomial). This is because all the syzygies of the second kind involve factors like $(\alpha-k)$, $k\in\mathbb{Z}$, so certain relations among bracket polynomials degenerate when α happens to be integral. These remarks will be treated in greater detail in a forthcoming paper.

4. Characterization of Homogeneous Divergences. The solution of problem 1 of the introduction using the transform proceeds as follows. First, for later reference, we determine the transform of a Jacobian determinant

$$J = \partial(u_{I_1},\dots,u_{I_r})/\partial(x^{k_1},\dots,x^{k_r}) = \det(\partial u_{I_\nu}/\partial x^{k_\mu})$$

of derivatives of u. A simple calculation shows that

$$\hat{J} = (r!)^{-1}\det(Z^K)\cdot\det(Z^{\mathcal{I}}),$$

where Z^K denotes the r×r minor consisting of <u>columns</u> $k_1,\dots,k_r$ of the matrix Z, and $Z^{\mathcal{I}}$ denotes the r×r matrix of monomials $(z_\mu)^{I_\nu}$ which occur in the symmetric powers of the matrix Z.

Theorem. Let q=1. Then every differentially homogeneous divergence Q = Div P is a linear combination of Jacobian determinants of derivatives of u.

Proof. Consider the transform $\hat{Q}$ of Q. Using the formula for the transform of the total derivatives, we see that Q is a divergence if and only if

$$\hat{Q}(Z) = \textstyle\sum_i (z_1^i + \dots + z_r^i)\cdot\hat{P}_i(Z)$$

for polynomials $\hat{P}_j$. This means that

$$\hat{Q}(Z) = \hat{Q}(z_1,\dots,z_r) = 0 \text{ whenever } z_1 + \dots + z_r = 0,$$

z_i denoting the i-th row of Z. Moreover, differential homogeneity of $\hat{Q}$ implies that Q is a homogeneous function of degree k of the rows of Z, so

$$\hat{Q}(\lambda_1 z_1,\dots,\lambda_r z_r) = (\lambda_1\cdot\ldots\cdot\lambda_r)^k\cdot\hat{Q}(z_1,\dots,z_r).$$

Thus we have the stronger condition

$$\hat{Q}(z_1,\dots,z_r) = 0 \text{ whenever } z_1,\dots,z_r \text{ are linearly dependent.}$$

If p<r, it is easy to see that there are no nontrivial such polynomials $\hat{Q}$. Otherwise, let $\mathcal{I}^*$ denote the determinantal ideal generated by the r×r minors of the r×p matrix Z. Then the above condition is equivalent to the fact that $\hat{Q}$ vanishes on the ideal $\mathcal{I}^*$. According to a theorem of Northcott, [13], and Mount, [12], $\mathcal{I}^*$ is a prime ideal, and so by the Hilbert Nullstellensatz

$$\hat{Q}(Z) = \textstyle\sum_K (\det Z^K)\cdot\tilde{R}_K(Z)$$

for some polynomials $\tilde{R}_K(Z)$. Finally, we use the fact that $\hat{Q}$ is symmetric; applying the symmetrizing map σ to the last formula, we can replace $\tilde{R}_K$ by its skew-symmetrization $\hat{R}_K$, which is easily seen to be a linear combination of the power determinants $\det(Z^{\mathcal{I}})$ occurring in the transform of the Jacobian determinants. Therefore $\hat{Q}$ is a linear combination of transforms of Jacobian determinants, and hence inverting the transform proves that Q itself is a linear combination of Jacobian determinants, which completes the proof of the theorem.

5. Hyperjacobians and Transvectants. The corresponding problem for homogeneous higher order divergences is approached in the same manner. We begin by transforming the basic condition

$$Q = \sum D_I P_I,$$

which gives the polynomial equation

$$\hat{Q}(Z) = \sum (z_1 + \ldots + z_r)^I \cdot \hat{P}_I(Z),$$

where $(z_1 + \ldots + z_r)^I$ denotes the I-th power of the sum of the rows of Z. Again, differential homogeneity of Q implies that $\hat{Q}$ is an algebraically homogeneous function of the rows of Z, so the above condition is equivalent to the condition that $\hat{Q}$ and all its partial derivatives up to order k-1 with respect to the variables z^i_j lie in the determinantal ideal $\mathcal{J}^*$. In other words, $\hat{Q}$ lies in the k-th <u>symbolic power</u> of $\mathcal{J}^*$.

At this point, we require an important result of Trung, [22], (see also DeConcini, Eisenbud and Procesi, [5]) that for maximal sized minors of a matrix of independent variables Z, the k-th symbolic power of the corresponding determinantal ideal $\mathcal{J}^*$ is the same as the ordinary k-th power of $\mathcal{J}^*$. Therefore, any such $\hat{Q}$ can be written as a sum of powers of determinants:

$$\hat{Q}(Z) = \sum (\det Z^{I_1}) \cdot \ldots \cdot (\det Z^{I_k}) \cdot \hat{R}^{\mathcal{J}}(Z).$$

Moreover, applying σ as before, we find that $\hat{R}^{\mathcal{J}}(Z)$ can be written as a linear combinations either of power determinants $\det(Z^{\mathcal{J}})$ if k is odd, or power permanants $\operatorname{perm}(Z^{\mathcal{J}})$ if k is even. Thus, to complete the characterization of homogeneous higher order divergences, we need to know which differential polynomials transform to

$(\det Z^{I_1}) \cdot \ldots \cdot (\det Z^{I_k}) \cdot (\det Z^{\mathcal{J}})$, k odd, or $(\det Z^{I_1}) \cdot \ldots \cdot (\det Z^{I_k}) \cdot (\operatorname{perm} Z^{\mathcal{J}})$, k even.

The answer is given by the theory of <u>hyperjacobians</u>. Rather than give the most general definition of a hyperjacobian, which is a generalization of the classical concept of a Jacobian determinant, we present the second degree examples, from which the general definition can easily be guessed.

Example. A first order hyperjacobian is simply a Jacobian determinant, e.g.

$$\partial(u,v)/\partial(x^1,x^2) = u_1 v_2 - u_2 v_1.$$

A second order hyperjacobian is obtained from a first order one by a similar determinantal formula:

$$\partial^2(u,v)/\partial(x^1,x^2)\partial(x^3,x^4) = \partial(u_3,v_4)/\partial(x^1,x^2) - \partial(u_4,v_3)/\partial(x^1,x^2)$$

$$= u_{13}v_{24} - u_{23}v_{14} - u_{14}v_{23} + u_{24}v_{13}.$$

In particular,

$$\partial^2(u,v)/\partial(x^1,x^2)^2 = u_{11}v_{22} - 2u_{12}v_{12} + u_{22}v_{11}$$

is a multiple of the second transvectant of u and v, and

$$\partial^2(u,u)/\partial(x^1,x^2)^2 = 2(u_{11}u_{22} - u_{12}^2)$$

agrees, up to a factor, with the Hessian of u. Similarly, third order hyperjacobians are constructed from second order hyperjacobians by the same determinantal procedure; for example

$$\partial^3(u,v)/\partial(x^1,x^2)^3 = \partial(u_1,v_2)/\partial(x^1,x^2)^2 - \partial(u_2,v_1)/\partial(x^1,x^2)^2$$
$$= u_{111}v_{222} - 3u_{112}v_{122} + 3u_{122}v_{112} - u_{222}v_{111}$$

is a multiple of the third transvectant of u and v, cf. [9; page 227]. In general, if u and v are homogeneous polynomials, the k-th transvectant $(u,v)^{(k)}$ is just a multiple of the k-th order hyperjacobian $\partial^k(u,v)/\partial(x^1,x^2)^k$.

The general formula for the hyperjacobian

$$\partial^k(u^1,\dots,u^r)/\partial(x^{i_1},\dots,x^{i_r})\dots\partial(x^{m_1},\dots,x^{m_r})$$

is found inductively on k, using a similar determinantal construction. They can also be constructed using the theory of higher dimensional determinants of higher order Jacobian "matrices", cf. [14]. They provide the natural generalization of the notion of transvectant to higher dimensional problems. Furthermore, their transforms are precisely what is required to complete the analysis of higher order divergences, so we have the following result:

Theorem. A differential polynomial Q is a k-th order divergence if and only if it is a linear combination of k-th order hyperjacobians of derivatives of u.

In particular, any k-th order transvectant is a k-th order divergence, a fact that does not appear to have previously been noticed in the literature.

6. Differential Hyperforms. Although we now know that any k-th order hyperjacobian can be written as a k-th order divergence, the determination of exactly how to do this is a non-trivial task. One solution to this algebraic problem has resulted in the development of a new theory of higher order differential forms, or "hyperforms". The motivation for this is the observation that the divergence identity for the ordinary Jacobian determinant is equivalent to the differential form identity

$$du \wedge dv = d(u \cdot dv).$$

The goal is to develop a theory of differential forms so that the identity

$$\partial^2(u,v)/\partial(x^1,x^2)^2 = D_1^2(-u_2v_2) + D_1D_2(u_1v_2+u_2v_1) + D_2^2(-u_1v_1)$$

for the second order hyperjacobian (second transvectant) translates into an identity of the form

$$d^2u*d^2v = d^2(du*dv)$$

for second order hyperforms. Such a thory has been developed over Euclidean space in the unpublished paper [15], and we here summarize its principal ingredients.

Let λ denote a Young diagram, or shape, and $|\lambda|$ the number of boxes in λ. Given a finite dimensional real vector space V, we let $L_\lambda V$ denote the corresponding irreducible representation space of the general linear group GL(V); L_λ is known as the Schur functor (or shape functor). See [1], [11], [21], for the general theory of Schur functors. Pieri's formula is

$$V \otimes L_\lambda V = \oplus_\mu L_\mu V,$$

where the direct sum is over all shapes μ containing λ and having precisely one box more than λ. For notation, we write $\lambda \subset \mu$ when λ is contained in μ, and write $\mu \backslash \lambda$ for the skew shape consisting of all boxes of μ which do not lie in λ; thus the above direct sum would be over all $\mu \supset \lambda$ with $|\mu \backslash \lambda| = 1$. This formula implies the existence of functorial maps

$$\varphi_\lambda^\mu : V \otimes L_\lambda V \to L_\mu V$$

for any such λ, μ, which are uniquely defined up to scalar multiple.

Example. Recall first that the Schur space $L_\lambda V$ can be identified with the quotient space of the tensor product of the symmetric powers $\odot_{\lambda_1}V \otimes \ldots \otimes \odot_{\lambda_k}V$ under the two-sided ideal generated by the Young relations

$$\sum x_{\hat{\imath}} \otimes (x_i \odot y),$$

for $x_1,\ldots,x_{p+1} \in V$, $y \in \odot_\ell V$, and where $x_{\hat{\imath}} = x_1 \odot \ldots \odot x_{i-1} \odot x_{i+1} \odot \ldots \odot x_{p+1}$. Therefore, each element ω of $L_\lambda V$ can be identified with an equivalence class in $\odot_{\lambda_1}V \otimes \ldots \otimes \odot_{\lambda_k}V$, and we can use any element of this equivalence class to represent ω. With this in mind we will write elements of $L_\lambda V$ as if they were in $\odot_{\lambda_1}V \otimes \ldots \otimes \odot_{\lambda_k}V$, but with the understanding that we are allowed to replace such a representative by any equivalent element as prescribed by the Young relations.

We can write down the explicit formulas for the Pieri maps φ^μ_λ. The easiest approach is to present a few special cases, from which the general formula can be deduced. First, if $\lambda=(k,\ell)$, $k\geq\ell$, and $\mu=(k+1,\ell)$, then

$$\varphi^\mu_\lambda(v\otimes(x\otimes y)) = (v\odot x)\otimes y + \tfrac{1}{k-\ell+2}\textstyle\sum_{i=1}^{\ell} (x\odot y_i)\otimes(v\odot y_{\hat{i}}).$$

If $\mu=(k,\ell+1)$ (so $k>\ell$),

$$\varphi^\mu_\lambda(v\otimes(x\otimes y)) = x\otimes(v\odot y),$$

while if $\mu=(k,\ell,1)$, then

$$\varphi^\mu_\lambda(v\otimes(x\otimes y)) = x\otimes y\otimes v.$$

Similarly, if $\lambda=(k,\ell,m)$, $k\geq\ell\geq m$, and $\mu=(k+1,\ell,m)$, then

$$\varphi^\mu_\lambda(v\otimes(x\otimes y\otimes z)) = (v\odot x)\otimes y\otimes z + \tfrac{1}{k-\ell+2}\textstyle\sum_{i=1}^{\ell} (x\odot y_i)\otimes(v\odot y_{\hat{i}})\otimes z +$$
$$+\tfrac{1}{k-m+3}\cdot\textstyle\sum_{j=1}^{m}(x\odot z_j)\otimes y\otimes(v\odot z_{\hat{j}}) + \tfrac{1}{k-\ell+2}\cdot\tfrac{1}{k-m+3}\sum_{i=1}^{\ell}\sum_{j=1}^{m} (x\odot y_i)\otimes(y_{\hat{i}}\odot z_j)\otimes(v\odot z_{\hat{j}}),$$

while if $\mu=(k,\ell+1,m)$, so $k>\ell$, then

$$\varphi^\mu_\lambda(v\otimes(x\otimes y\otimes z)) = x\otimes(v\odot y)\otimes z + \tfrac{1}{\ell-m+2}\textstyle\sum_{j=1}^{m} x\otimes(y\odot z_j)\otimes(v\odot z_{\hat{j}}).$$

Also, for $\mu=(k,\ell,m+1)$ (so $\ell>m$) and $\mu=(k,\ell,m,1)$ we have

$$\varphi^\mu_\lambda(v\otimes(x\otimes y\otimes z)) = x\otimes y\otimes(v\odot z), \quad \text{and} \quad \varphi^\mu_\lambda(v\otimes(x\otimes y\otimes z)) = x\otimes y\otimes z\otimes v$$

respectively. The general formula appears in [15].

In any of the above formulas, we could multiply φ^μ_λ by a constant c^μ_λ, without affecting the functoriality. It turns out that one can choose constants c^μ_λ in such a way that the resulting Pieri products

$$v*\omega = c^\mu_\lambda\cdot\varphi^\mu_\lambda(v\otimes\omega), \quad v\in V, \quad \omega\in L_\lambda V, \quad v*\omega\in L_\mu V,$$

commute in the sense that $v*v*\omega$ is unambiguously defined as an element of $L_\nu V$ whenever $\lambda\subset\nu$ and $|\nu\setminus\lambda| = 2$, in which case the two $*$ products can be computed in two distinct ways, depending on which box is added on to λ first. One possible choice of the factors c^μ_λ is the following: if $\lambda=(\lambda_1,\dots,\lambda_n)$, and $\mu=(\lambda_1,\dots,\lambda_{j-1},\lambda_j+1,\lambda_{j+1},\dots,\lambda_n)$, then set

$$c^\mu_\lambda = \begin{cases} (\lambda_1+n)^{-1}, & j=1,\\ (\lambda_1-\lambda_j+j-1)\cdot\prod_{k=j+1}^{n+1}(\lambda_j-\lambda_k+k-j)^{-1}, & 1<j\leq n,\\ 1 & j=n+1,\end{cases}$$

where $\lambda_{n+1}=0$ by convention. For instance, if $\lambda=(k,\ell)$, $\mu=(k+1,\ell)$, then $c^\mu_\lambda = 1/(k+2)$, while if $\mu=(k,\ell+1)$, $c^\mu_\lambda = (k-\ell+1)/(\ell+1)$. Finally, let v be a fixed nonzero vector in V, and define the map $\psi^\mu_\lambda\colon L_\lambda V\to L_\mu V$ by

$$\psi^\mu_\lambda(\omega) = v*\omega = c^\mu_\lambda\cdot\varphi^\mu_\lambda(v\otimes\omega), \quad \omega\in L_\lambda V.$$

Lemma. Let $\lambda\subset\mu\subset\nu$, and $\lambda\subset\mu'\subset\nu$ be shapes, with $|\mu\setminus\lambda|=1$, $|\mu'\setminus\lambda|=1$, and $|\nu\setminus\mu|=1$, $|\nu\setminus\mu'|=1$ (so $|\nu\setminus\lambda|=2$). Then

$$\psi^\nu_\lambda = \psi^\nu_\mu\circ\psi^\mu_\lambda = \psi^\nu_{\mu'}\circ\psi^{\mu'}_\lambda$$

commute and serve to unambiguously define a map ψ_λ^ν from $L_\lambda V$ to $L_\nu V$.

By the lemma, it is easy to see that, given $v \in V$, we can uniquely define a map $\psi_\lambda^\mu: L_\lambda V \to L_\mu V$ for any shape μ containing λ; if $|\mu \backslash \lambda| = m$, then we define ψ_λ^μ by m-fold composition of the basic Pieri maps, adding on one box at a time. The lemma assures us that it does not matter in which order the boxes are added on.

Theorem. Let $0 \neq v \in V$, and let $\psi_\lambda^\mu: L_\lambda V \to L_\mu V$, $\lambda \subset \mu$, be the corresponding maps of Schur spaces. Then the ψ_λ^μ define a <u>exact hypercomplex</u> over V. Here "hypercomplex", which is a generalization of the usual homological algebraic concept of a complex, means that these maps satisfy the two properties of

a) <u>Commutativity</u>: Whenever $\lambda \subset \mu$ and $\mu \subset \nu$, then $\psi_\lambda^\nu = \psi_\mu^\nu \circ \psi_\lambda^\mu$.

b) <u>Closure</u>: Whenever $\lambda \subset \nu$ and $\nu \backslash \lambda$ contains two or more boxes in any one column, then $\psi_\lambda^\nu = 0$.

Furthermore, the Schur hypercomplex is exact, meaning that

c) <u>Exactness</u>: Suppose $\lambda \subset \mu$ and $\mu \subset \nu$, and $\mu \backslash \lambda$ and $\nu \backslash \mu$ each consist of only one row of boxes, and $\nu \backslash \lambda$, which consists of two rows of boxes, has two boxes in one and only one column. Then, $\psi_\mu^\nu(\omega) = 0$ for $\omega \in L_\mu V$ if and only if $\omega = \psi_\lambda^\mu(\theta)$ for some $\theta \in L_\lambda V$.

(There is a more general statement of exactness which characterizes the kernel of ψ_μ^ν for any $\mu \subset \nu$, but this is a little harder to state, and the above notion of exactness is sufficient for this more general version to hold; see [15] for details.)

Contained in the Schur hypercomplex, corresponding to the shapes with only one column of boxes, is the standard exterior complex $\omega \mapsto v \wedge \omega$, $\omega \in \Lambda_k V$, and in this case, closure and exactness reduce to their more familiar counterparts.

What we are really interested in is the differential counterpart of the Schur hypercomplex. At present, the construction has only been carried out for open subsets M of Euclidean space $\mathbb{R}^p$, but extensions to arbitrary smooth manifolds are possible, once certain technical details regarding changes of varialbe have been overcome. Let T^*M be the cotangent bundle of M, with $T^*M|_x$, which will play the role of our vector space V, denoting the cotangent space of M at $x \in M$. Let $\Xi_\lambda|_x = L_\lambda T^*M|_x$ be the corresponding Schur space at x, with Ξ_λ the corresponding "Schur bundle", constructed in the exact same manner as the exterior powers $\wedge_k T^*M$, which are just special cases when λ has only one column of boxes. A <u>differential hyperform of shape λ</u> will be a section of Ξ_λ. More explicitly, let $dx^1,\dots,dx^p$ be the standard basis of T^*M. Then corresponding to every Young tableau of shape λ whose entries are chosen from $\{1,\dots,p\}$, there is an element dx_T of Ξ_λ. The elements dx_T corresponding to standard tableaux form a basis of Ξ_λ, cf. [1]. (Here standard means that the rows of the tableau are nondecreasing and the columns strictly increasing.) Therefore any differential hyperform can be written in the form

$$\omega = \sum f_T(x)\cdot dx_T,$$

where the sum is over all standard tableaux of shape λ. If $\lambda\subset\mu$ and μ differs from λ by only one box, then we define the <u>differential</u> of ω to be the differential hyperform

$$d\omega = d^\mu_\lambda\omega = \sum_T df_T(x) * dx_T = \sum_T c^\mu_\lambda\cdot\varphi^\mu_\lambda(df_T(x)\otimes dx_T),$$

where $df_T \in T^*M$ denotes the ordinary differential of the coefficient function f_T. The algebraic commutativity lemma immediately implies the commutativity of the differentials: Let $\lambda\subset\mu\subset\nu$, and $\lambda\subset\mu'\subset\nu$ be shapes, with $|\mu\setminus\lambda|=1$, $|\mu'\setminus\lambda|=1$, and $|\nu\setminus\mu|=1$, $|\nu\setminus\mu'|=1$ (so $|\nu\setminus\lambda|=2$). Then $d^\nu_\mu\circ d^\mu_\lambda = d^\nu_{\mu'}\circ d^{\mu'}_\lambda$ commute and serve to unambiguously define a second order differential d^ν_λ from Ξ_λ to Ξ_ν. Therefore, by composition, we can define a differential $d = d^\mu_\lambda\colon \Xi_\lambda \to \Xi_\mu$ whenever $\lambda\subset\mu$. The main result from [15] is:

Theorem. Let $M \subset \mathbb{R}^p$ be a star-shaped domain. Then the <u>differential hypercomplex</u> defined by the differentials $d^\mu_\lambda\colon \Xi_\lambda \to \Xi_\mu$ on hyperforms is an exact hypercomplex.

Contained within the differential hypercomplex, corresponding to those shapes with only one column, is the ordinary deRham complex over M, where closure and exactness reduce to their usual meanings. There are also a host of other interesting subcomplexes of the full hypercomplex, but space precludes us presenting this in any detail here. If M is not star-shaped, then one can use the failure of the differential hypercomplex to be exact to define new "hypercohomology" over M, but this has not been investigated at all.

If $\lambda=0$ is the "empty shape", then $\Xi_\lambda|_x = \mathbb{R}$, and a "0-hyperform" is just an ordinary real-valued function. If $\mu = (k)$ consists of a single row, so $\Xi_\mu = \odot_k T^*M$ is the k-th symmetric power of T^*M, then $d_0^\mu u = d^k u$ is just the k-th differential of the function u, i.e.

$$d^k u = \sum_I \binom{k}{I} u_I \cdot dx_I,$$

the sum being over all multi-indices I of order k, with u_I denoting the corresponding k-th order partial derivative. One immediate consequence of the exactness of the differential hypercomplex is that, provided M is star-shaped, a section ω of $\Xi_\mu = \odot_k T^*M$ is a k-th order differential, $\omega = d^k f$ for some function f, if and only if $d_\mu^\nu \omega = 0$, where $\nu = (k,1)$ is the two-rowed shape with one box in the second row.

Turning to the hyperjacobian identities, let μ be the rectangular shape consisting of p rows, each consisting of k boxes, i.e. $\mu = (k,k,\dots,k)$. It can be shown, using the Littlewood-Richardson rule, that the representation space $L_\mu V$ occurs with multiplicity one in the p-fold tensor product $\otimes_p \odot_k V$ of the k-th symmetric power of V. Hence there is, up to constant multiple, just one functorial map $\pi\colon \otimes_p \odot_k V \to L_\mu V$, and that is given by identifying $d^k u^1 \otimes \dots \otimes d^k u^p \in \otimes_p \odot_k V$ as an element of $L_\mu V$. A straight-forward calculation using the straightening rule shows that the coefficient of $(dx^1)^k \otimes \dots \otimes (dx^p)^k$ is the k-th order hyperjacobian $\partial^k(u^1,\dots,u^p)/\partial(x^1,\dots,x^p)^k$. (This can be generalized to other hyperjacobians by changing the number of u's.) For example, as an element of Ξ_μ, where $\mu = (2,2)$,

$$\begin{aligned} d^2u \otimes d^2v = \\ &= \{u_{11}(dx^1)^2 + 2u_{12}dx^1 \odot dx^2 + u_{22}(dx^2)^2\} \otimes \{u_{11}(dx^1)^2 + 2u_{12}dx^1 \odot dx^2 + u_{22}(dx^2)^2\} \\ &= \{u_{11}v_{22} - 2u_{12}v_{12} + u_{22}v_{11}\}\,(dx^1)^2 \otimes (dx^2)^2 \end{aligned}$$

since

$$(dx^1)^2 \otimes (dx^2)^2 + 2(dx^1 \odot dx^2) \otimes (dx^1 \odot dx^2) = 0$$

by the Young relation.

Furthermore, let λ consist of p-1 rows with k boxes in each row. Let $k = sp + t$, where $0 \le t < p$. Then $L_\lambda V$ occurs with multiplicity one in the tensor product $\otimes_t \odot_{k-s-1} V \otimes \otimes_{p-t} \odot_{k-s} V$, and so, up to constant multiple, there is a unique functorial map $\hat{\pi}\colon \otimes_t \odot_{k-s-1} V \otimes \otimes_{p-t} \odot_{k-s} V \to L_\lambda V$. (This is a little harder to describe explicitly; see [15].) Fixing this multiple appropriately, we find that the hyperform identity

$$d^k u^1 \otimes \dots \otimes d^k u^p = d^k(d^{k-s-1}u^1 \otimes \dots \otimes d^{k-s-1}u^t \otimes d^{k-s}u^{t+1} \otimes \dots \otimes d^{k-s}u^p)$$

is equivalent to the expression of the above k-th order hyperjacobian as a k-th order divergence. Here we are identifying the hyperform in parenthesis with its image in Ξ_λ under the map $\hat{\pi}$, and $d^k = d_\lambda^\mu$.

Example. For the Hessian, we need to look at the identity

$$d^2u\otimes d^2v = d^2(du\otimes dv),$$

where $d^2u\otimes d^2v \in \Xi_{(2,2)}$, and $du\otimes dv \in \Xi_{(2)}$. We find

$$du\otimes dv = -\tfrac{1}{2}du\odot dv = -\tfrac{1}{2}\{u_1v_1(dx^1)^2 + (u_2v_1+u_1v_2)dx^1\odot dx^2 + u_2v_2(dx^2)^2\}$$

and

$$d^2(du\otimes dv) = -\tfrac{1}{2}\{d^2(u_1v_1)*(dx^1)^2 + d^2(u_2v_1+u_1v_2)*dx^1\odot dx^2 + d^2(u_2v_2)*(dx^2)^2\}$$
$$= \{-D_2^2(u_1v_1) + D_1D_2(u_2v_1+u_1v_2) - D_1^2(u_2v_2)\}\cdot(dx^1)^2\otimes(dx^2)^2,$$

and we recover the second order transvectant identity. See [14], [15], for further identities of this type.

References

[1] K. Akin, D. A. Buchsbaum and J. Weyman, "Schur functors and Schur complexes", Adv. in Math. 44 (1982), 207-278.

[2] Ball, J. M., Currie, J.C. and Olver, P.J., "Null Lagrangians, weak continuity, and variational problems of arbitrary order", J. Func. Anal. 41 (1981), 135-174.

[3] Coleman, B.D. and Noll, W., "The thermodynamics of elastic materials with heat conduction and viscosity", Arch. Rat. Mech. Anal. 13 (1963) , 167-178.

[4] Cushman, R. and Sanders, J.A., "Nilpotent normal forms and representation theory of sl(2,ℝ)", Univ. of Amsterdam, Report #301, 1985.

[5] DeConcini, C., Eisenbud, D. and Procesi, C., "Young diagrams and determinantal varieties", Invent. Math. 56 (1980), 129-165.

[6] Dunn, J.E. and Serrin, J., "On the thermomechanics of interstitial working", Arch. Rat. Mech. Anal. 88 (1985), 95-133.

[7] Gel'fand, I.M. and Dikii, L.A., "Asymptotic behaviour of the resolvent of Sturm-Liouville equations and the algebra of the Korteweg-deVries equations", Russ. Math. Surveys 30 (1975), 77-113.

[8] Grace, J.H. and Young, A., The Algebra of Invariants, Cambridge Univ. Press, Cambridge, 1903.

[9] Gurevich, G.B., Foundations of the Theory of Algebraic Invariants, P. Noordhoff Ltd., Groningen, Holland, 1964.

[10] Kung, J.P.S. and Rota, G.-C., "The invariant theory of binary forms", Bull. Amer. Math. Soc. 10 (1984), 27-85.

[11] Lascoux, A. "Syzygies des varietés determinantales", Adv. in Math. 30 (1978), 202-237.

[12] Mount, K.R., "A remark on determinantal loci", J. London Math. Soc. 42 (1967), 595-598.

[13] Northcott, D.G., "Some remarks on the theory of ideals defined by matrices", Quart. J. Math. Oxford 65 (1963), 193-204.

[14] Olver, P.J., "Hyperjacobians, determinantal ideals and weak solutions to variational problems", Proc. Roy. Soc. Edinburgh 95A (1983), 317-340.

[15] Olver, P.J., "Differential hyperforms I", Univ. of Minn. Math. Report 82-101, 1983.

[16] Olver, P.J., "Conservation laws and null divergences", Math. Proc. Camb. Phil. Soc. 94 (1983), 529-540.

[17] Olver, P.J., "Conservation laws and null divergences II. Nonnegative divergences", Math. Proc. Camb. Phil. Soc. 97 (1985), 511-514.

[18] Olver, P.J., Applications of Lie Groups to Differential Equations, Graduate Texts in Mathematics, vol. 107, Springer-Verlag, New York, 1986.

[19] Shakiban, C., "A resolution of the Euler operator II", Math. Proc. Camb Phil. Soc. 89 (1981), 501-510.

[20] Shakiban, C. "An invariant theoretic characterization of conservation laws", Amer. J. Math. 104 (1982), 1127-1152.

[21] Towber, J., "Two new functors from modules to algebras", J. Algebra 47 (1977), 80-104.

[22] Trung, N.V., "On the symbolic powers of determinantal ideals", J. Algebra 58 (1979), 361-379.

Computing Invariants*

George R. Kempf

The Johns Hopkins University

Let G be a reductive group over an algebraically closed field k of characteristic zero. If $\mathbb{V}$ is a representation of G on an affine space, one considers the ring $k[\mathbb{V}]^G$ of polynomial functions on $\mathbb{V}$ which are invariant under the action of G. Then the ring $k[\mathbb{V}]^G$ of invariants is a graded ring with finite dimensional graded pieces. In the classical case, where $G = SL_n$ and $\mathbb{V}$ is space of homogeneous polynomials of degree d in n variables, Hilbert [2] proved the ring of invariants was a finitely generated k-algebra. This wonderful theorem just raised the question of finding an algorithm for computing these generators. Hilbert [3] gave a difficult but finite procedure for finding the generators in his case.

Since Hilbert's time some progress has taken place. H. Weyl [10] proved the finiteness of generators for arbitrary reductive groups, which are exactly the groups with compact real forms on which his unitary trick worked. D. Mumford [8] generalized Hilbert numerical criterion for instability which as we shall see is an essential for a good understanding of the null cone. M. Hochster and J. Roberts [4] discovered a theorem, that $k[\mathbb{V}]^G$ is a Cohen-Macaulay ring, which was not even conjectured by the classical workers. An outgrowth of their cohomological work will provide an improvement of the last step of the algorithm.

In this paper I will give an exposition of the subjects mentioned above. Then I will give an exposition of the present form of Hilbert's idea of an algorithm for computing invariants. This algorithm uses some commutative

*Partly supported by NSF Grant #MPS75-05578

algebra of polynomial rings which now is on the level of computer programable calculations (anyway it can be done effectively in a finite number of steps).

Lastly I will present a new result which gives a large estimate on the maximal degree of generating invariants. This a priori estimate is the outgrowth of work at P. Gordon, E. Noether and V. Popov.

§1. The generating invariants

We will denote the ring $k[V]$ of polynomials on V by A and the ring of invariants $k[V]^G$ by I. Following the classical terminology an invariant will be a homogeneous element of I of positive degree. Let I^+ be the positively graded part of I. Then I^+ is the graded ideal generated by the invariants.

The reductive algebraic groups are exactly those with completely reducible representations. Therefore there is a unique G-invariant k-linear operator $U : A \longrightarrow I$ which projects A onto its invariant subspace I. It follows easily that $U(ai) = U(a)i$ and $i = U(i)$ for all elements i of I and all elements a of A. In other words U is an I-module retraction of the inclusion $I \subset A$.

Now we can prove the original theorem of invariant theory.

Theorem 1. The ring I is finitely generated over k.

Proof. Consider the ideal (I^+) in A generated by all invariants. By the basis theorem we may find a finite set J of invariants such that $(J) = (I^+)$. Let i be an invariant. Then $i \in I^+$. So we may write $i = \sum_{j \in J} a_j \cdot j$ where a_j in A. Apply U to this equation and letting b_j be the invariant $U(a_j)$, we have $i = \sum_{j \in J} b_j \cdot j$ where the invariants b_j have lower degree than i. By an induction on degree, any invariant is a polynomial in the j's. Therefore the finite set J generates I.

The problem with finding such a finite set J is that we do not know the ideal (I^+) of A. It will turn out that the radical of (I^+) is more accessible. Let N denote the closed subset zeroes$\{I^+\}$ of $\backslash V$. As (I^+) is a homogeneous ideal, N is a cone. This cone N is called the null cone as it is the locus where all the invariants are zero.

We may find a finite set K of invariants such that $N = \text{zeroes}\{K\}$. Then we have an improvement of the last result.

Theorem 2. The ring I is a finitely generated module over its sub-k-algebra generated by the finite set K. Conversely if I is integral over the sub-ring generated by invariant say L, then $N = \text{zeroes}\{L\}$.

Proof. By the nullstellen satz, $(I^+)^p \subset (K)$ for some positive integer p. Let d be a maximal degree of the generator in J. Clearly any invariant i of degree $\geq d\cdot p$ is contained in $(I^+)^p$ and, hence, in (K). Thus we may write $i = \sum_{k\varepsilon K} a_k\cdot k$ with a_k in A. As in the last proof we have $i = \sum_{k\varepsilon K} b_k\cdot k$ where the invariants b_k are of lower degree than i. This time our induction shows that $i = \sum c_p\cdot d_p$ where the c_p's are invariants of degree $< d\cdot p$ and the d_p's are polynomials in the elements of K. As we may choice the c_p's from a finite k-basis of invariants of such bounded degree, the first statement is true.

As for the last statement, by assumption for any invariant i satisfies a non-trivial equation $i^n + b_1 i^{n-1} + \ldots + b_n = 0$ of minimal degree where the b_i's are polynomials in the elements of L. By homogeneity, the b_i's have no constant terms. Hence they vanish at any point in zeroes$\{L\}$. Consequently i must vanish there. Thus $N \equiv \text{zeroes}\{I^+\} \supset \text{zeroes}\{L\}$. As the reverse inclusion is obvious, we have $N = \text{zeroes}\{L\}$. QED

A slight improvement of the last result is

Theorem 3. We may choice the elements of K to be algebraically independent over k.

Proof. This argument of Hilbert's is now called the normalization theorem and I will only sketch the proof. Replacing the invariants K of Theorem 2 by appropriate powers, we may assume that they have the same degree. Then we may find a finite set L of linear combinations of K so that the elements of L are algebraically independent and the k-algebra generated by K in integral over the polynomial ring $k[L]$. By the converse in Theorem 2, it follows that L is a possible choice of K. QED

From the algorithmic point of view one may pass from a set K as in Theorem 2 to one satisfying Theorem 3 in a finite number of steps but the maximal degree of the elements of the new $K \leq$ least common multiple of the degree of all elements of the old K.

We will call a set K of invariants which satisfies Theorem 3 a set of primary invariants.

§2. The geometry of the null cone

Hilbert discovered how to give a geometric description of the null cone N which needs no knowledge of invariant theory. We will review this shortly but first we will see how this applies to computing invariants.

Assuming that the null cone N is known, we may examine all invariants of higher and higher degree. Eventually we will find a finite set K of invariant such that $N = \text{zeroes}\{K\}$. Then we may apply Theorems 2 and 3. Thus after a finite number of steps we may find a finite set of primary invariants.

The first clue to giving a concrete description of the null cone is contained in the following definition. A vector v of $\mathbb{V}$ is called G-unstable if 0 is contained in the closure $\overline{G\,v}$ of its orbit under G. With this notion we have a geometric description of N.

Theorem 4. The null cone N consists precisely of the G-unstable vectors of $\mathbb{V}$.

Proof. Any invariant has constant value on $\overline{G\cdot v}$. Hence if v is G-unstable, then any invariant vanishes at v because it vanishes at zero. So the null cone contains the G-unstable vectors.

Assume that v is not G-unstable. Then 0 and $\overline{G\cdot v}$ are disjoint G-invariant closed subsets. The function f with value 0 on 0 and 1 on $\overline{G\cdot v}$ is a regular function on $0 \amalg \overline{G\cdot v}$. By complete reduciblity we may lift the G-invariant function f to an element i of A. As i has no constant term, there must be an invariant which does not vanish on $\overline{G\cdot v}$. Hence the null cone does not contain any vectors which are not G-unstable. QED

Hilbert had two ideas how to use this result. The first was generalized by V. Popov [9] for the case of a semi-simple connected group. The second idea has led to interesting generalizations by D. Mumford and his school to various numerical criterions for stability in geometric invariant theory. (See the appendices to the second edition of Geometric Invariant Theory [8].) I will give a quick exposition of the original version of Hilbert.

A one-parameter subgroup λ of G is a homomorphism $\lambda : G_m = \{t \in k | t \neq 0\} \longrightarrow G$. The heart of the numerical criterion for instability is

Theorem 5. A vector v of $\mathbb{V}$ is contained in the null cone N if and only if there is a one-parameter subgroup λ of G such that $\lim_{t\to 0} \lambda(t)\cdot v = 0$.

Proof. Clearly if the limit is zero, 0 is contained in the closure $\overline{G\cdot v}$. Hence v is contained in N by Theorem 4. The converse may be found in [8] or [5]. QED

The reason that the word numerical is used in the following. Let $v \neq 0$ be a vector in $\mathbb{V}$. We have an eigendecomposition $v = \sum v_i$ under the action of λ. Thus $\lambda(t)\cdot v = \sum t^i v_i$. Consider Mumford's number $m_\lambda(v)$ = minimum i such that $v_i \neq 0$. Then $\lim_{t\to 0} \lambda(t)\cdot v = 0$ if and only if $m_\lambda(v) > 0$.

Let T be a maximal torus of G. Let C denote the null cone for T acting on $\mathbb{V}$. Following Hilbert we call C the canonical cone. Next we will see that any null vector is conjugate by G to one in canonical form (i.e. contained in C).

<u>Theorem 6.</u> $N = G\cdot C$.

<u>Proof.</u> If $\lim_{t\to 0} \lambda(t)\cdot v = 0$ for some vector v and some one-parameter subgroup λ of G, we may find an element of G such that $\lambda' = g\lambda g^{-1}$ is contained in T as all maximal torus of G are conjugate to T. Hence $\lim_{t\to 0} \lambda'(t)\cdot(gv) = 0$ and hence $g\cdot v$ is contained in C. Thus by Theorem 5 $N \subset G.C$. As the reverse inclusion is obvious we are done. QED.

The canonical cone C is very computable. It is the union of a finite number of linear subspaces of V which are determined by the eigenspace decomposition of V.

Let $\mathbb{V} = \bigoplus_{\chi\in\Theta} V_\chi$ where $0 \neq V_\chi = \{v \in V | t\cdot v = \chi(t)\sigma$ for all t in $T\}$

Θ is a finite set of characters of T.

The set P of one-parameter subgroup of T form a free abelian group of rank equal to the dimension of T. We may think of P as a lattice in the real vector space $P \otimes_{\mathbb{Z}} \mathbb{R} \equiv Q$. We have a perfect pairing of P with the group of characters of T given by $\langle\lambda,\chi\rangle = n$ where $\chi(\lambda(t)) = t^n$. We may identify characters with linear function on Q.

For a one-parameter subgroup λ of T, let

$$\mathbb{V}(\lambda) = \{v \in \mathbb{V} \mid \lim_{t\to 0} \lambda(t)\cdot v = 0\} = \bigoplus_{\langle\lambda,\chi\rangle>0} V_\chi .$$

Let H be the set of connected components of $Q - \bigcup_{\chi\in\Theta}\{\langle -,\chi\rangle = 0\}$. For any component h in H $\mathbb{V}(\lambda)$ is independent of the choice of λ in $h \cap P$. Let this subspace be denoted by $\mathbb{V}_h$. Furthermore for any one-parameter subgroup λ of T, $\mathbb{V}(\lambda)$ is contained in $\mathbb{V}_h$ if λ is contained in the closure $\bar{h}$ of h. Thus by the numerical criterion for T-instability we get

<u>Lemma 7</u> $\qquad C = \bigcup_{h\in H} \mathbb{V}_h$

This completes the description of the null cone N.

§3. <u>The modern ideas</u>

Let J be a primary system of invariants. The remarkable result of Hochster-Robert [4, 6] is

<u>Theorem 8</u>. There exist a finite set S of secondary invariants such that I is the free $k[J]$-module with basis $S' = \{1\} \cup S'$.

In other words we may write any invariant uniquely in the form $\sum_{s\in S'} f_s \cdot s$ where f_s is a polynomial in the primary invariants of J.

The invariants in $S' = \{s_0 = 1, s_1, \ldots, s_r\}$ may be found inductively. Just take s_ℓ to be an invariant of smallest degree in $I - \sum_{0\le i\le \ell-1} k[J]s_i$. This process has a definite end by

<u>Theorem 9</u>. The maximum degree of any secondary invariant is less than or equal to the sum of the degrees of the primary invariants [6].

With this last theorem we see that there is a finite procedure for finding all invariants. These modern developments are fortunate because I do not understand how to extend Hilbert's last argument to an arbitrary reductive group. Even more fortunately the Theorems 3 and 9 are quantitative. This allows us to prove

Corollary 10. Let h = max degree (k) in a set K as in Theorem 2. Then the ring I is generated by invariants of degree $\leq$ (dim $\mathbb{V}$) $C(h)$ where $C(h)$ = least common multiple of the numbers $\leq h$.

Proof. By the proof of Theorem 3 we know that we may find a primary system J of invariants of degree $\leq C(h)$. By Theorem 9 the degree of the secondary invariants $\leq \#(J)\cdot C(h)$ but $\#(J) = \dim I \leq \dim \mathbb{V}$. QED

Example. Let $\mathbb{G}_m$ act on $\mathbb{A}^3$ by the formula $t(x,y,z) = (t^3x, t^{-2}y, t^{-5}z)$. Then the null cone is $N = (x=0) \cup (y=z=0)$. The monomials $x^2y^3 = j_1$ and $x^5y^3 = j_2$ are algebraically independent. Also $N = \{j_1 = j_2 = 0\}$. Thus j_1 and j_2 are primary invariants. Now $13 = 5+8 = \deg j_1 + \deg j_2$ is the upper bound for the degrees of the secondary invariants. One finds all invariants of degree ≤ 13. They are j_1, j_1^2, j_2, j_1j_2, $s_1 = x^3y^2z, j_1s_1$, $s_2 = x^4yz^2$ and s_2j_1. Thus s_1 and s_2 are secondary invariants. Hence I is generated by j_1, j_2, s_1 and s_2 which all have degree ≤ 8.

For the last result we have a very strong form of finite generation.

Theorem 11. There is an easy calculation given G and $\mathbb{V}$ which produces a number $N_{G,\mathbb{V}}$ such that I is generated by invariants of degree $\leq N_{G,\mathbb{V}}$.

Proof. We will use two previously known special cases.

1) (E. Noether [7]) If G is finite, we can take $N_{G,\mathbb{V}} = \#G$.

2) (V. Popov [9]) If G is semi-simple and connected and the representation of G on $\mathbb{V}$ is almost faithful; i.e. if the homomorphism $G \to GL(\mathbb{V})$ is finite, we can take

$$N_{G,\mathbb{V}} = n\, C\left(\frac{2^{r+s} n^{s+1} (n-1)^{s-r} t^r (s+1)!}{3^s \left(\left(\frac{s-r}{2}\right)!\right)^2}\right)$$

where $n = \dim \mathbb{V}$, $s = \dim G$, r = rank of G and t is a number (to be explained) which measure the complicity of the eigenfunction of a maximal torus T of G

acting on $\mathbb{V}$. Explicitly take an isomorphism of T with the r-fold product of $\mathbb{G}_m$'s. Let χ be a character of T. Then $\chi(t_1,\ldots,t_r) = \pi t_i^{m_i}$ for some integers $m_1,\ldots,m_r$. Let $||\chi|| = \max|m_i|$. Then $t = \max ||\chi||$ where χ runs through the eigenfunction for T acting on $\mathbb{V}$. Actually Popov states his result for a faithful representation but his argument works equally well for an almost faithful one.

The new special case will be proven in the next section.

3) If G is torus, we may take

$$N_{G,\mathbb{V}} = n\ C(n\ s!\ w^s)$$

where $n = \dim \mathbb{V}$, $s = \dim G$ and w is a number similar to t; i.e., take an isomorphism $G = G_m{}^s$ then $w = \max ||\chi||$ for eigenfunctions χ in $\mathbb{V}$ where $||\chi||$ is defined above.

Next I will explain how these special cases imply the theorem.

Assume that G is a normal subgroup of another reductive group H such that the quotient H/G is reductive. Also assume that we have a representation of H on $\mathbb{V}$ which extends the action of G. Then we have an induced action of H on $I = k[\mathbb{V}]^G$ as G is normal. This action induces an action of H/G on I. Let $\mathbb{W}$ be the affine space with linear functions $I \cap k[\mathbb{V}]_{1\le i\le N_{G,\mathbb{V}}}$ where $k[\mathbb{V}]_{1\le i\le N_{G,\mathbb{V}}}$ denotes the sum of homogeneous terms of degree between 1 and $N_{g,\mathbb{V}}$. Then $\mathbb{W}$ is a representation of H/G. By complete reducibility, the homomorphism

$$k[\mathbb{W}]^{H/G} \longrightarrow [k[\mathbb{V}]^G]^{H/G} = k[\mathbb{V}]^H$$

is surjective. Therefore we may take

$$N_{H,\mathbb{V}} = N_{H/G,\mathbb{W}} \cdot N_{G,\mathbb{V}}\ .$$

This allows one to find $N_{H,\mathbb{V}}$ by working with the smaller groups G and H/G.

Consider the pair of groups, G^0=connected component of G and G itself. The quotient G/G^0 is finite. By the above remark and step 1) we may reduce to the case where G is connected.

Next let S be the radical of G. Then S is a normal subgroup which is a torus and the quotient G/S is connected semi-simple. To do this case we need to know 3) and also 2). If the representation of G/S on $\mathbb{W}$ is almost faithful, to use 2) we need to have an estimate for $\dim \mathbb{W}$ and the number t for the action of G/S on $\mathbb{W}$. To do this first note that

$\dim \mathbb{W} \leq \dim k[\mathbb{V}]_{1\leq i\leq N_{S,\mathbb{V}}}$ which is a well known function of $\dim \mathbb{V}$ and $N_{S,\mathbb{W}}$. It remains to bound t. Let T be a maximal torus of G. Then T/S is a maximal torus of G/S. There is an integer z such that we have a commutative diagram

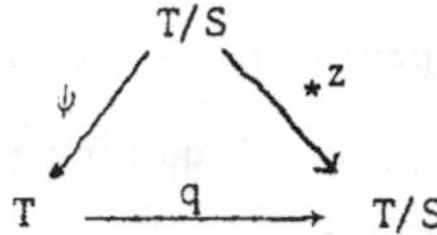

where q is the quotient homomorphism and $*^z$ is the z-power mapping. Now take an isomorphism $T/S \overset{\sim}{\sim} X\mathbb{G}_m$. Clearly the number t for T/S acting on $\mathbb{W}$ is less than or equal to $\dfrac{(t')^{N_{S,V}}}{k}$ where t' is the similar number for the action of T/S on $\mathbb{V}$ which is induced by ψ.

This works if the representation of G/S on $\mathbb{W}$ is almost faithful but we must consider the other cases also. In any case we have two normal subgroups H_1 and H_2 of G/S such that $G/S = H_1 \cdot H_2$ and H_1 acts almost faithfully of $\mathbb{W}$ and H_2 acts trivially on H_2. Then H_1 is connected semi-simple and has maximal torus $T/S \cap H_1$. We may apply the above method to estimate $N_{H_1,\mathbb{W}}$ as there are only finitely many possibilities for H_1. Therefore it remains to prove 3.

§4. The toroidal case

We have a representation $\mathbb{V}$ of a torus T. We will fix an isomorphism $T = \mathbb{G}_m^s$ and $\mathbb{V} = \mathbb{A}^n$ such that the standard basis of $\mathbb{A}^n$ consists of eigenvectors. Thus $(t_1,\dots,t_s)(x_1,\dots,x_n) = (\pi_j t_j^{m_{ij}} x_i)$ where m_{ij} is a matrix of integers where $1 \le i \le n$ and $1 \le j \le s$. The number $w = \max |m_{ij}|$.

By the previous theory we want to prove that

a) there is a set K of invariant with degree $\le ns!w^s$ such that all invariants are linearly dependent on them.

This will establish the estimate 3) by Corollary 10.

We will explicitly construct the set K. Any invariant is a linear combination of invariant monomials $\pi_{1 \le i \le n} x_i^{\ell_i}$. The condition that this is an invariant monomial is

b) $\sum_{1 \le i \le n} \ell_i m_{ij} = 0$ for all $1 \le j \le s$ and

$\ell_i \in \mathbb{N}$ for all $1 \le i \le n$.

This system of integral conditions b) was originally considered by P. Gordon [1]. He proved the finite generation of the ring of invariants by showing that the semi-group of solution of b) is finitely generated.

We want to find a finite (minimal) set J of solutions of b) such that

I) any solution i of b) satisfies

$$h_i = \sum_{j \in J} r_j\, j$$

for some positive integer h and non-negative integers r_j.

This means that the h power of the invariant monomial corresponding to i is a product of invariant $\tilde{J}$ corresponding to J. Hence any invariant is integrally dependent on $\tilde{J}$. To bound the degree we need

II. $\deg(j) \equiv \sum_{1 \le i \le n} j_i \le ns!w^s$ for all j in J

or more strongly

II'. for all $j = (j_i)$ in J, $|j_i| \leq s!\,w^s$ for all $1 \leq i \leq n$.

Then I and II' imply a) with $K = \mathfrak{J}$.

Let X be the rational vector subspace of $\mathbb{Q}^n$ generated by the vectors (m_{ij}) for $1 \leq j \leq s$. Let y_i be the image in $Y = \mathbb{Q}^n/X$ of the i-th coordinate vector x_i in $\mathbb{Q}^n$. Let $R \subset [1,\ldots,n]$ be any set such that $\{y_i\}_{i\in R}$ form a basis for a hyperplane in Y. Then there are precisely two non-zero integral vector (ℓ_i) in $\mathbb{Z}^n$ with relative prime coefficients which satisfy

c) $\left| \begin{array}{ll} \sum_{1\leq i\leq n} \ell_i m_{ij} = 0 & \text{for all } 1\leq j\leq s \text{ and} \\ \ell_k = 0 & \text{for all } k \text{ in } R. \end{array} \right.$

Let J be the set of all such vectors for all possible R which also satisfy $\ell_i \geq 0$ for all $1 \leq i \leq n$. Clearly the elements of J satisfy b). We need to check I) and II') are true.

The condition II' follows from elementary linear algebra. Let $S \subset [1,\ldots,s]$ be a subset such that (m_{ij}) for j in S form a basis for X. Then $\#S + \#R = n-1$ for any possible choice of R. We identify the dual space of $\mathbb{Q}^n$ with $\Lambda^{n-1}(\mathbb{Q}^n)$ as usual. Then the vector $\ell = \bigwedge_{j\in S}(m_{ij}) \quad \bigwedge_{r\in R} x_r$ is a solution of c) with integral coefficients. Computing the coefficients of ℓ by determinants, we find that they are determinants of $\#S\times\#S$ minor of (m_{ij}). As $|m_{ij}| < w$, any coefficient of $\ell \leq (\#S)!\,w^{\#S} \leq s!\,w^s$. As our corresponding element of J has the form $\pm\frac{1}{c}(\ell)$ where c is the greatest common divisor of the coefficients of ℓ, II' is true.

The proof of I is more geometric. Let $C = \sum_{1\leq i\leq n} \mathbb{Q}_{>0} \cdot y_i$ be the rational polyhedral cone in Y spanned by the y_i's. Considering the inclusion $Y^\wedge \subset (\mathbb{Q}^n)^\wedge$ of dual vector spaces, we may regard the set of all solutions of b) as the set of integral points of the dual cone $C^\wedge = \{\ell \in Y^\wedge \mid \langle \ell, c\rangle \geq 0$ for all c in C}. This dual cone $C^\wedge$ is another rational polyhedral cone. A one-dimensional face $\mathbb{Q}_{\geq 0}\cdot m$ of $C^\wedge$ is dual to a $n-\#S-1$ dimensional faces M of C and the hyperplane $m = 0$ is spanned by M. As such a hyperplane is spanned by a set $\{y_i\}_{i\varepsilon R}$ for some R, we see that J consists of the minimal integral vectors of the one dimensional faces of $C^\wedge$. To prove that I is true we will show that

I') $$C^\wedge = \sum_{j\varepsilon J} \mathbb{Q}_{\geq 0}\cdot j .$$

In other words $C^\wedge$ is spanned by its one-dimensional spaces.

The proof of I' will use the real numbers.

Let $C_{\mathbb{R}} \subset Y_{\mathbb{R}}$ be the cone $\sum_{1\leq i\leq n} \mathbb{R}_{\geq 0}\cdot y_i$. Then we have the dual cone $C_{\mathbb{R}}^\wedge$ is $Y_{\mathbb{R}}^\wedge$. We first note that $C_{\mathbb{R}}^\wedge = \sum_{j\varepsilon J} \mathbb{R}_{\geq 0}\cdot j$. To see this note that $C_{\mathbb{R}}$ contains an open subset of $Y_{\mathbb{R}}$ as the y_i's span Y. Let d be a point of this open set. Then $(\langle -,d\rangle=0) \cap C_{\mathbb{R}}^\wedge$ contains only zero. Let S be a sphere in $Y_{\mathbb{R}}$ around the origin. Then $S \cap C_{\mathbb{R}}^\wedge$ is compact and contained in the half-plane $\langle -,d\rangle > 0$. Clearly $S \cap C_{\mathbb{R}}^\wedge$ is homeomorphic to the convex set $E \equiv C_{\mathbb{R}}^\wedge \cap (\langle -,d\rangle=1)$. Thus E is compact convex polyhedron. By the Krein-Milman theorem E is the convex hull of its vertices which are $\frac{1}{\langle j,d\rangle}\cdot j$ where j runs through J. This means that $C_{\mathbb{R}}^\wedge = \sum_{j\varepsilon J} \mathbb{R}_{\geq 0}\cdot j$.

As $C^\wedge = C_{\mathbb{R}}^\wedge \cap Y^\wedge$, we need to see that $(\sum_{j\varepsilon J} \mathbb{R}_{\geq 0}\, j) \cap Y^\wedge = \sum_{j\varepsilon J} \mathbb{Q}_{\geq 0}\, j$. This will prove I'). Let L be a line in $Y^\wedge$. We want to see that if $(\sum_{j\varepsilon J} \mathbb{R}_{\geq 0}\cdot j) \cap L \neq 0$, then L contains a non-zero element of $\sum_{j\varepsilon J} \mathbb{Q}_{\geq 0}\cdot j$. Consider the abstract rational vector space Z with basis $\underline{j}$ where j in J. There is a linear mapping $\psi : Z \longrightarrow Y^\wedge$ which sends $\underline{j}$ to j. By assumption $\psi^{-1}(L) - \psi^{-1}(0)$ meets $\sum_{j\varepsilon J} \mathbb{R}_{\geq 0}\,\underline{j}$ in $Z_{\mathbb{R}}$ in some point and, hence, it meets this simplicial cone in the interior of some face $\sum_{j\varepsilon S} \mathbb{R}_{\geq 0}\,\underline{j}$ for some $S \subset J$.

Then $\psi^{-1}(L) \cap \sum_{j\in S} \mathbb{Q}_{\geq 0}\, \underline{j}$ is a rational polyhedron which "contains" an open subset of real points not contained in $\psi^{-1}(0)$. By density of rational points $\psi^{-1}(L) - \psi^{-1}(0)$ contains a point $\sum_{j\in S} a_j \underline{j}$ where the a_j's are non-zero rational numbers. Thus $\sum_{j\in S} a_j \cdot j$ is the required element of L. This ends our proof.

References

1. P. Gordon, Vorlesungen über Invariantentheorie, vol. 1, p. 199, Teubner, Leipzig, 1885.
2. D. Hilbert, Über die Theorie der algebraischen Formen, Math. Ann. 36, 1890, p. 473-534.
3. ———, Über die vollen Invariantensysteme, Math. Ann. 42, 1893, p. 313-373.
4. M. Hochster and J. Roberts, Rings of invariants are Cohen-Macaulay, Adv. in Math, 13, 1974, p. 115-175.
5. G. Kempf, Instability in invariant theory, Ann. of Math, 108 (1978) p. 299-316.
6. ———, The Hochster-Roberts theorem of invariant theory, Mich. Math. Jour. 26, 1979, p. 19-32.
7. E. Noether, Math. Ann. 77, 1916, p. 89-92.
8. D. Mumford, Geometric Invariant Theory, Erg. der Math 34, Springer, 1982.
9. V. Popov, Constructive Invariant Theory, p. 303-334, Asterisque 87-88.
10. H. Weyl, Theorie der Darstellung kontinuierlicher halbeinfacher Gruppen durch lineare Transformationen, Math. Zeit. 24 (1926), p. 377-395.

CONSTRUCTING INVARIANT POLYNOMIALS
via
TSCHIRNHAUS TRANSFORMATIONS

Frank D. Grosshans
Department of Mathematical Sciences
West Chester University of Pennsylvania
West Chester, Pennsylvania 19383/USA

The technique of using Tschirnhaus transformations to generate invariants was used successfully in the theory of binary quantics. These transformations led to systems of semi-invariants, called protomorphs, which then were used to construct invariants, themselves [2;Chapter χ]

In this chapter, we extend this method of protomorphs to semi-simple algebraic groups defined over an algebraically closed field k. The basic theorem is proved in (1.4). This theorem is then applied to the construction of invariants in Section 2. In Section 3, we review the classical example.

1. Tschirnhaus Transformations

(1.1) Weights and roots. Let G be a connected semi-simple algebraic group. Let $B = TU$ be a Borel subgroup of G. Let π be the system of simple roots corresponding to U and let Δ be the set of all positive roots. There is an element w_s in the Weyl group of T which sends π to $-\pi$.

Let ω be the highest weight of an irreducible representation of G acting on a vector space V and let $\omega_o = w_s \cdot \omega$. Let

$$\Delta' = \{\alpha \in \Delta : \langle\omega_o,\alpha\rangle \neq 0\}.$$

Then $\langle\omega_o,\alpha\rangle < 0$ for all $\alpha \in \Delta'$ [10;Theorem 39, p.209].

Let α be any root of T with respect to U. We shall denote by $x_\alpha(c)$, $c \in k$, the corresponding one-dimensional, unipotent subgroup of G.

Lemma [10; Lemma 72,p.209]. Let v be a vector in V having weight λ. There are vectors v_i in V having weights $\lambda + i\,\alpha$ $(i=1,2,\ldots)$ so that $x_\alpha(c) \cdot v = v + \Sigma c^i v_i$ for all $c \in k$.

We shall use an extension of this lemma which may be proved by induction. Let v be a vector in V having weight λ and let $\alpha_1,\ldots,\alpha_m$ be any m roots in Δ. Then, for any $c_1,\ldots,c_m$ in k, we have

(*) $$x_{\alpha_1}(c_1) \cdots x_{\alpha_m}(c_m) \cdot v$$

is a linear combination of vectors having weights $\lambda + e_1\alpha_1 + \ldots + e_m\alpha_m$ where $e_1,\ldots,e_m$ are non-negative integers.

In particular, let v_o be a vector in V having weight ω_o. Then $V = kUv_o$ [10;Theorem 39,p.209]. It follows from (*), then, that each weight in V has the form $\omega_o + e_1\alpha_1 + \ldots + e_m\alpha_m$.

(1.2) <u>Assumption</u>. Let v_o be a vector in V having weight ω_o and let α be a root in Δ'. According to the lemma in (1.1),

$$x_\alpha(c) \cdot v_o = v_o + \Sigma c^i v_i$$

where v_i has weight $\omega_o + i\alpha$. In this expression, we shall assume that v_1, the vector of weight $\omega_o + \alpha$, is not zero. And, we shall denote v_1 by v_α.

In case k has characteristic 0, the assumption always holds. If k has characteristic $p > 0$, then the assumption holds if p does not divide $<\omega_o,\alpha>$. This can be proved by considering the precise description of the irreducible representation of SL_2 given in [9;pp.44-45], for example.

(1.3) <u>Basis of weight vectors</u>. From now on, let us put $\Delta' = \{\alpha_1,\ldots,\alpha_m\}$. Let v_o be a vector in V having weight ω_o. Let $v_i = v_{\alpha_i}, i=1,\ldots,m$, be vectors in V, as in (1.2), having weights $\omega_o + \alpha_i$. We extend $\{v_o,v_1,\ldots,v_m\}$ to a basis for V, $\{v_o,v_1,\ldots,v_m,\ldots,v_n\}$, which consists of weight vectors of T. Let us denote the corresponding dual basis by $\{\mu_o,\mu_1,\ldots,\mu_n\}$.

(1.4) <u>Theorem</u>. Let $V' = \{v \in V : \mu_o(v) \neq 0\}$; let $W = \{v \in V : \mu_i(v) = 0$ for $i=1,\ldots,m\}$ and let $W' = W \cap V'$. Let

$$U' = \Pi\, x_\alpha(c_\alpha),\ \alpha \in \Delta' .$$

(i) Then, U' is the unipotent radical of a parabolic subgroup of G.

(ii) The mapping $U' \times W' \to V'$ given by $(u,w) \to u \cdot w$ is an isomorphism whose inverse will be denoted by $v \to (\phi(v)^{-1}, \phi(v) \cdot v)$.

<u>Proof</u>. To prove (i), let $P = \{g \in G : gv_o = cv_o\}$. Then P is a parabolic subgroup of G corresponding to a subset $-\pi_o$ of $-\pi$. The unipotent radical of P is generated by those $x_\beta(c)$, $\beta \in -\Delta$, such that $<\omega_o,\beta> > 0$. Let P_1 be the parabolic subgroup of G corresponding to the subset π_o of π. The unipotent radical of P_1 corresponds to those roots $\alpha \in \Delta$ so that $<\omega_o,\alpha> < 0$. Hence, this unipotent radical is U'.

To prove (ii), we first show that the mapping $U' \times W' \to V'$ is one-to-one. Let $u \in U'$ and $w \in W'$ be the elements so that $u \cdot w \in W'$. We will prove that $u = e$. Let

$$u = \Pi\, x_\alpha(c_\alpha), \quad \alpha \in \Delta'.$$

We will show that $c_\alpha = 0$ by an induction argument on $ht\alpha$ (Let $\pi = \{\beta_1,\ldots,\beta_\ell\}$; then $\alpha = m_1\beta_1 +\ldots+ m_\ell\beta_\ell$ and we recall that $ht\alpha = m_1 +\ldots+ m_\ell$.) If $ht\alpha = 1$, then α is simple and the coefficient of v_α in $u \cdot w$ is $c_\alpha\mu_o(w)$. But $u \cdot w$ is in W' so $c_\alpha = 0$. Now suppose that $ht\alpha = h$ and that $c_\beta = 0$ for all β in Δ' satisfying $ht\beta < h$. Then, using (*) in (1.1), we see that the coefficient of v_α in $u \cdot w$ is $c_\alpha\mu_o(w)$. Hence, $c_\alpha = 0$.

To complete the proof, we will display an everywhere defined mapping $\phi : V' \to U'$ so that $\phi(v) \cdot v \in W'$ for all $v \in V'$. By induction on $ht\alpha$, we will define functions $c_\alpha(v) \in k[V']$ so that

$$\phi(v) = \Pi\, x_\alpha(c_\alpha(v)), \quad \alpha \in \Delta'.$$

If $ht\alpha = 1$, then the coefficient of v_α in $\phi(v) \cdot v$ would be $\mu_o(v)c_\alpha + \mu_\alpha(v)$. Hence, we set

$$c_\alpha(v) = -\mu_\alpha(v)/\mu_o(v).$$

For the induction step, let us suppose that for each β in Δ' such that $ht\beta < ht\alpha$ we have defined functions $c_\beta(v)$ in $k[V']$ which satisfy the following condition: if

$$u_o = \Pi x_\beta(c_\beta(v)), \quad \text{then} \quad \mu_\beta(u_o \cdot v) = 0.$$

The coefficient of v_α in $\phi(v) \cdot v$ would then be

$$\mu_o(v)c_\alpha + \mu_\alpha(v) + (\text{a polynomial in the } c_\beta\text{'s}, \mu_\beta(v), \mu_o(v)^{-1}).$$

Therefore, we may solve for c_α.

(1.5) Properties of ϕ . The mapping $\phi: V' \to U'$ defined in (1.4) has several properties which we will use later.

P1) $\phi(cv) = \phi(v)$ for all $v \in V'$, $c \in k$, $c \neq 0$.
P2) $\phi(uv) = \phi(v)u^{-1}$ for all $v \in V'$, $u \in U'$.
P3) $\phi(tv) = t\phi(v)t^{-1}$ for all $v \in V'$, $t \in T$.

These properties follow immediate from the following fact: If $v \in V'$, then $\phi(v)$ is the unique element in U' such that $\phi(v) \cdot v$ is in W'.

2. Invariant Functions

(2.1) U'-invariant functions. The group U' operates on $U' \times W'$ via $u \cdot (u_1,w) = (uu_1,w)$. Each orbit contains one and only one point of the form (e,w). Hence, the U'-invariant functions on $U' \times W'$ can be identified with $k[W']$. Since the isomorphism from $U' \times W' \to V'$ is U'-equivariant, we obtain the following

Theorem. The mapping $\psi : V' \to W'$ given by $\psi(v) = \phi(v) \cdot v$ gives rise to an isomorphism

$$\psi^o : k[W'] \to k[V']^{U'}$$

defined by $\psi^o(f)(v) = f(\psi(v)) = f(\phi(v) \cdot v)$. Furthermore, $(\psi^o)^{-1}f = f|W'$, the restriction of f to W'.

Now, we have $k[W'] = k[\mu_o,\mu_{m+1},\ldots,\mu_n][1/\mu_o]$. Let us put

$$\mu_i(\phi(v) \cdot v) = S_i(v)/\mu_o(v)^{e_i}, \ i=m+1,\ldots,n,$$

where $S_i \in k[V]$ and e_i is a non-negative integer. Then

$$k[V']^{U'} = k[\mu_o,S_{m+1},\ldots,S_n][1/\mu_o].$$

Corollary 1. The orbits of U' on V' are closed in V and are separated by $\mu_o,S_{m+1},\ldots,S_n$.

Proof. The orbits of U' on V' are closed and separated by the functions above since a corresponding statement holds on $U' \times W'$. Also, the orbits of U' on V are all closed [8;Theorem 2,p.221].

Corollary 2. The U'-invariant functions on V consist of all those functions in $k[\mu_o,S_{m+1},\ldots,S_n][1/\mu_o] \cap k[V]$

or

$$k[V]^{U'} = \{\psi^o f : f \in k[W], \ \psi^o f \in k[V]\}.$$

Note 1. If $\text{char}\, k \geq 0$, then $k[V]^{U'}$ is known to be finitely generated [4].

Note 2. Let H be a subgroup of U'. A slight extension of the argument above shows that

$$k[V']^H = k[U'/H] \otimes k[W'].$$

If H is normalized by T, then $k[U'/H]$ is a polynomial algebra.

Note 3. The mapping $\psi : V' \to W'$ gives a geometric quotient of V' by U' in the sense of [6;p.70]

(2.2) Observable hull of TU'. Let G be an affine algebraic group. A closed subgroup H of G is called observable if there is a finite-dimensional vector space V on which G acts and a vector v in V so that $H = \{g \in G : g \cdot v = v\}$.

Given any subgroup H of G, there is a smallest observable subgroup of G containing H; this has been denoted by H'' [3;p.236]. We shall call H'' the observable hull of H. The importance of this concept in invariant theory is the following result [3;Lemma 1,p.242]: if Z is an affine variety on which G operates, then $k[Z]^H = k[Z]^{H''}$.

Returning to the language of Section 1, we will investigate the G-invariant functions on V by first showing that G is the observable hull of TU' if G acts on V with finite kernel. This result depends heavily on the theorems proved in [7]. I am indebted to K. Pommerening for the Theorem, below, and its proof and, also, for the proof of Theorem 1,(2.3).

Theorem. Let G be a connected semi-simple algebraic group. Let U' be the unipotent radical of a parabolic subgroup $\mathbf{P}$ having maximal torus T. If $\mathbf{P}$ does not contain an infinite normal subgroup of G, then the observable hull of TU' is G.

Proof. Let Φ and Ψ be the root systems of G and U', respectively. The observable hull H of TU' has $\Psi \cup (-\Psi)$ in its root system and is the smallest reductive subgroup containing TU'. Let I be the subset of the corresponding root system that defines the reductive part of the parabolic subgroup. We need to show that each root α in I belongs to H. Such an α is in a connected component of I in the Dynkin diagram and this component is connected to a basis root β in Ψ. Let $\alpha = \alpha_0, \alpha_1, \ldots, \alpha_m, \alpha_{m+1} = \beta$ be the connecting way. Then the root $\beta' = \sigma_{\alpha_1} \cdots \sigma_{\alpha_m} \beta$ is in Ψ and $\langle\beta', \alpha\rangle < 0$. Hence, α and β' form an irreducible root system of $\Phi' = \Phi \cap (\mathbf{Z}\alpha + \mathbf{Z}\beta')$ having rank two. Without loss of generality, we may assume that $\Phi = \Phi'$ with basis $\alpha \in I$, $\beta \in \Psi$. Then $\alpha' = \sigma_\beta \alpha$ is in Ψ and $\beta' = \sigma_\beta \beta = -\beta$ is in $(-\Psi)$ and form a basis, too. Chevalley's commutator formulas for root systems of rank 2 [5;Section 33] shows that the quasi-closure of $\Psi \cup (-\Psi)$ is Φ. This completes the proof.

(2.3) G-invariant functions. We keep the notation used in (1.1). Let $\omega_1, \ldots, \omega_\ell$ be a fundamental system of highest weights for T with respect to the order imposed by U. Let γ be a one-parameter subgroup of T so that $\langle\omega_i, \gamma\rangle > 0$ for each $i=1,\ldots,\ell$.

Theorem 1. Let G act on V with finite kernel. Then the observable hull of $\gamma \cdot U'$ is G.

Proof. This follows at once from [7;(3.6)]. Indeed, (3.6) (i) is in (3.5), Example 2 and (3.6) (ii) was shown in the proof of the Theorem in (2.2) above.

Theorem 2. Let G act on V with finite kernel. Then

$$k[V]^G = \{\psi^o f : f \in k[W]^\gamma, \psi^o f \in k[V]\}.$$

Proof. Let $f \in k[W]^\gamma$ satisfy $\psi^o f \in k[V]$. Then $\psi^o f$ is fixed by U' according to Corollary 2 in (2.1) and by γ according to P3) in (1.5). Therefore, $\psi^o f$ is fixed by G according to Theorem 1, above.

Note 1. The algebra $k[W]^\gamma$ can be described explicitly [8;p.217].

Note 2. Let f be a homogeneous polynomial in $k[W]^\gamma$ and let $f = \Sigma c_i M_i$ where the M_i are γ-invariant monomials. The condition that $\psi^o f$ be in $k[V]$ defines certain linear relations on the c_i. Furthermore, if $\psi^o f$ is in $k[V]$, then $\deg f = \deg \psi^o f$. (For the polynomials S_i defined above are homogeneous of degree $e_i + 1$ according to P1.)

3. The Example of $SL_2(\mathbb{C})$

(3.1) The action. The usual action of $SL_2(\mathbb{C})$ on $\mathbb{C}^2$ (viewed as 2×1 matrices) gives rise to an action on polynomials in the variables x and y, namely,

$$\begin{pmatrix} a & b \\ c & d \end{pmatrix} \cdot x = dx-by \text{ and } \begin{pmatrix} a & b \\ c & d \end{pmatrix} \cdot y = -cx + ay.$$

We consider the action of $SL_2(\mathbb{C})$ on the space of polynomials in x and y which are homogeneous of degree p. The notation is simplified somewhat if we identify $(a_0,a_1,\ldots,a_p)$ with

$$F(x,y) = a_0x^p + pa_1x^{p-1}y + \ldots + \binom{p}{i} a_ix^{p-i}y^i + \ldots + a_py^p.$$

(3.2) Tschirnhaus transformations. Let $W = \{(a_0,a_1,\ldots,a_p) : a_1 = 0\}$. If $F = (a_0,a_1,\ldots,a_p)$ with $a_0 \neq 0$, then

$$\phi(F) = \begin{pmatrix} 1 & a_1/a_0 \\ 0 & 1 \end{pmatrix}$$

and

$$\psi(F) = \phi(F) \cdot F = (a_0, 0, S_2(F)/a_0, \ldots, S_p(F)/a_0^{p-1})$$

where

$$\begin{aligned} S_2 &= a_0a_2 - a_1^2 \\ S_3 &= a_0^2a_3 - 3a_0a_1a_2 + 2a_1^3 \\ S_4 &= a_0^3a_4 - 4a_0^2a_1a_3 + 6a_0a_1^2a_2 - 3a_1^4 \end{aligned}$$

and so on.

(3.3) Invariant functions. The arguments in (2.1) may be applied now to see that

$$\mathbb{C}[V']^U = \mathbb{C}[a_0, S_2, \ldots, S_p][1/a_0].$$

The $SL_2(\mathbb{C})$ invariant functions consist of those f in $\mathbb{C}[W]^T$ such that $\psi^o f$ is everywhere defined.

Theorem. Let M be a monomial of degree d in $\mathbb{C}[W]$, say,

$$M = a_0^{e_0} a_2^{e_2} \ldots a_p^{e_p}.$$

(i) Then M is in $\mathbb{C}[W]^T$ if and only if $2e_2 + \ldots + pe_p = pd/2$.

(ii) If M is in $\mathbb{C}[W]^T$, then $\psi^o M$ has a pole at $a_0 = 0$ of order $d(p-2)/2$.

Proof. Condition (i) follows from considering the action of

$$t = \begin{pmatrix} c & 0 \\ 0 & 1/c \end{pmatrix}.$$

Next, $\psi^o M = a_0^{e_0}(S_2/a_0)^{e_2} \ldots (S_p/a_0^{p-1})^{e_p}$ has a pole at $a_0 = 0$ of order $e_2 + 2e_3 + \ldots + (p-1)e_p - e_0$. Adding $e_0 + e_2 + \ldots + e_p = d$, we see that

$$\text{(order of pole)} + d = 2e_2 + \ldots + pe_p = pd/2.$$

This proves (ii).

References.

1. A. Borel, Linear Algebraic Groups, notes by H. Bass. New York: W.A. Benjamin 1969.

2. E. Elliott, An Introduction to the Algebra of Quantics. 2nd ed. reprinted. New York: Chelsea Publishing Company 1964.

3. F. Grosshans, Observable groups and Hilbert's fourteenth problem. Amer. J. Math. 95, 229-253 (1973).

4. F. Grosshans, The invariants of unipotent radicals of parabolic subgroups. Invent. math. 73, 1-9 (1983).

5. J.E. Humphreys, Linear Algebraic Groups. New York: Springer-Verlag 1975.

6. P.E. Newstead, Introduction to Moduli Problems and Orbit Spaces. Bombay: Tata Institute 1978.

7. K. Pommerening, Observable radizielle Untergruppen von Halbeinfachen algebraischen Gruppen. Math. Zeitschrift. 165, 243-250 (1979).

8. M. Rosenlicht, On quotient varieties and the affine embedding of certain homogeneous spaces. Trans. Amer. Math. Soc. 101, 211-223 (1961).

9. T.A. Springer, Invariant Theory. New York: Springer-Verlag 1977.

10. R. Steinberg, Lectures on Chevalley Groups, Mimeographed lecture notes. New Haven: Yale Univ. Math. Dept. 1968.